EVOLUTION BY ASSOCIATION

A History of Symbiosis

JAN SAPP
Department of Science Studies
York University
Ontario, Canada

New York Oxford
OXFORD UNIVERSITY PRESS
1994

Oxford University Press

Oxford New York Toronto
Delhi Bombay Calcutta Madras Karachi
Kuala Lumpur Singapore Hong Kong Tokyo
Nairobi Dar es Salaam Cape Town
Melbourne Auckland Madrid

and associated companies in
Berlin Ibadan

Published by Oxford University Press, Inc.,
200 Madison Avenue, New York, New York 10016

Library of Congress Cataloging-in-Publication Data
Sapp, Jan.
Evolution by association : a history of symbiosis /
by Jan Sapp.
p. cm.
Includes bibliographical references and index
ISBN 0-19-508820-4 (cloth)—ISBN 0-19-508821-2 (paper)
1. Symbiosis—Research—History.
2. Symbiogenesis—Research—History.
I. Title.
QH548.S26 1994 574.5′2482′09—dc20 94-6321

2 4 6 8 9 7 5 3 1

Printed in the United States of America
on acid-free paper

FOR ELLIOT

Acknowledgments

This work could not have been completed without help from many people. I have profited greatly from interviews and discussions with biologists who introduced me to their field of research and ways of seeing. It is a pleasure to thank Ford Doolittle, William Eberhard, Michael Gray, Joshua Lederberg, Pedro Leon, David Lewis, Gabriel Macaya, Lynn Margulis, Paul Nardon, Jeremy Pickett-Heaps, David Read, David Smith, John Maynard Smith, Robert Trench, and Max Taylor.

Many others have offered comments and suggestions, and/or sent materials to me in the course of this research: Daniel Alexander, Linnda Caporael, Margalith Galun, Holger Gerberding, Elihu Gerson, Christiane Groeben, Ralph Lewis, Camille Limoges, Donna Mehos, Carole McKinnon, Gregg Mitman, Mary Beth Saffo, H. E. A. Schenk, Werner Schwemmler, Sorin Sonea, and William Summers.

Jacqueline Carpine Lancre, head librarian at the Muséum Océanographique de Monaco, provided invaluable assistance with the archives of Paul Portier, Jules Richard, and Prince Albert I of Monaco. I am also delighted to thank Paul Nardon for his help and encouragement, and for his hospitality in Lyon. Dimitri Krylov translated one of Famintsyn's Russian papers, and Henriette Gerberding translated one of his German papers.

I thank Lynn Margulis also for inviting me to the conference on symbiosis as a source of evolutionary innovation in Lake Como, Italy, in 1989, and for her hospitality in Amherst, Massachusetts, where my family and I spent the first half of 1991.

A rough draft of this book was completed during my subsequent year-long visit in Joshua Lederberg's laboratory at the Rockefeller University. I am grateful to Mary Jane Zimmerman and Joice Johnson for their kind assistance, and to my colleagues in the laboratory for providing a stimulating atmosphere in which to work.

Preliminary research for this book was carried out at the University of Melbourne with support from the Australian Research Grant Scheme. The Andrew W. Mellon Foundation and the Eugene Garfield Foundation supported my stay at the Rockefeller University. Subsequent support at York

University was generously provided by the Social Sciences and Humanities Research Council of Canada.

Finally, I thank Carole McKinnon and our two sons, Will and Elliot, for their endless inspiration.

Contents

Introduction

Revolutions in science have been declared throughout our century. One speaks of the revolution in quantum theory and the theory of relativity, the revolution in the earth sciences around continental drift and plate tectonics, the Mendelian revolution, the revolution in molecular biology. Each of these has been declared to have generated a transformation so broad and so deep that it touches everyone's most intimate sense of the nature of things.

Beginning in the 1960s, in the midst of the triumph of molecular biology, a "quiet revolution" in our understanding of ourselves took place. Our evolution, and that of all plants and animals, is not due solely to the gradual accumulation of gene changes within species. In fact, we evolved from, and are comprised of, a merger of two or more different kinds of organisms living together. Symbiosis is at the very root of our being. Inside each of our cells are tiny bodies called mitochondria, which are responsible for supplying most of the energy we need. Mitochondria have their own DNA, distinct from that of the chromosomes in the cell nucleus; they replicate and are transmitted from generation to generation maternally through the egg. They are now generally held to have originated as symbiotic bacteria some two billion years ago.

Chloroplasts, the energy-generating organelles of all plants, are also derived from independent, self-replicating lodgers: cyanobacteria. Like mitochondria, chloroplasts have their own DNA, RNA, and ribosomes. Similar ideas have been proposed for the origin of cilia and centrioles, the cell organelles concerned with motility and cell division. If these ideas are correct, the cells in our bodies would represent a synthesis of at least three different beings.[1] Contemporary symbiosis researchers are further urging evolutionists to reconsider other forms of symbiosis as representing a general macromechanism for the origin of species.[2]

Historians have analyzed the ramifications of Darwinism and metaphors of natural selection and "the struggle for existence" in both technical and social contexts. They have examined the "eclipse of Darwinism" during the first decades of the twentieth century, the union of Darwinism and Mendelian genetics, and the construction of the "evolutionary synthesis" of the 1930s and 1940s—based on gene mutation, recombination, and selection theory. In all of their rich and detailed analyses of evolutionary debates, symbiosis as a source of evolutionary novelty is virtually never mentioned.[3] Symbiosis is not included in the historical literature devoted to neo-Lamarckism and the inheritance of acquired characteristics; nor can one find it in historical writings on evolutionists who advocated a role for macromutations in the origin of species. The role of symbiosis in evolutionary change is also absent in historical studies about mutual aid[4] and cooperation in ecology.[5]

This absence reflects the diffidence about symbiosis on the part of biologists themselves. Yet the view that symbiosis is a major source of evolutionary novelty paralleled the development of Darwinian theory—if only as a dissenting footnote to that history. The suggestion was made the moment the term symbiosis was introduced in the late 1870s. The possibility that all plant and animal cells had evolved from symbiotic associations of two or more different organisms in the remote past was discussed by several individuals soon after. That symbiosis could lead to evolutionary change and the construction of new "individuals" was inferred from associations of algae and fungi in the formation of lichens, of photosynthetic algae in "lower animals," of nitrogen-fixing bacteria in the root nodules of legumes, and of fungi in the roots of forest trees and orchid bulbs. From the late nineteenth to the late twentith century, the scope and significance of such relationships were continuously, or rather discontinuously, debated.

Discussions were framed by two widely divergent perspectives. At one extreme was the view that such cases were illustrative of evolution in action: how "higher," more complex organisms could have evolved from "lower," simpler ones. Accordingly, there was a range of intimate relations between species, from parasitism to those in which associates mutually furthered and supported one another, and a continuum of dependency from transient to permanent interdependence. At the other extreme, many biologists insisted that stable and permanent interspecies integration rarely occurred; it was an exception to normal life, a curiosity that only detracted from proper biological aims and interests; and interpretations in terms of mutual assistance could be dismissed as mere sentimentalism.

The writings of contemporary symbiosis reseachers are replete with episodes and anecdotes about how their intellectual ancestors were ignored and ridiculed by an unappreciative scientific community. They emphasized how in Germany in the late nineteenth century, Andreas Schimper and Richard Altmann respectively proposed that chloroplasts and mitochondria were symbiotic lodgers inside cells. In Russia during the first decades of the twentieth century, the idea that chloroplasts were symbionts was developed by Andrei Famintsyn and by Konstantine Merezhkovskii, who coined the neologism "symbiogenesis" for the synthesis of new organisms from symbiotic unions. In France, Paul Portier's development of the idea that symbiosis was a universal phenomenon culminated in his book *Les Symbiotes* (1918), in which he argued that symbionts played an essential role in the physiology and development of all organisms and that mitochondria had originated as symbiotic bacteria. In the United States, Ivan E. Wallin developed similar ideas in

Symbionticism and the Origin of Species (1927). But it was not until the 1960s and 1970s, with the discovery that cell organelles possessed their own DNAs, that such ideas were taken seriously by mainstream biology.

Questing for a past and unfamiliar sensibility opens one to the temptation of taming shadowy and shifting forms to accommodate familiar shapes. Contemporary symbiosis researchers have often taken the research and ideas of past writers out of their own context, selectively citing those aspects which conform to the current conceptual concensus, and imposing new meaning on them (while still other writers with similar ideas have not been mentioned).

My object is not to construct a history of the research and ideas pertaining to symbiosis in terms of a progressive line or branching tree of ancestors; nor is it simply to trace out a hidden tradition around symbiogenesis—statements, beliefs, and rules that were handed down from generation to generation by word of mouth, if not by practice. It is to locate experimental studies of symbiosis within the life sciences ever since Darwin; to situate symbiosis within the complex matrix of biological specialization; to offer a broad sketch of the phenomena, movements, debates, doctrines, and metaphors that have shaped ideas and attitudes about its scope and significance.

Though few individual readers would deny that they know what symbiosis means, one would find no consensus among biologists as a group. While many equate symbiosis with mutual benefit of associated species, others have continuously argued that this narrow meaning of the term is difficult, if not impossible to apply to real associations. Some have applied the term to relations between individuals; others insist that it be restricted to interspecies relations; still others have limited it further to apply only to relations between species which remain in physical contact throughout most or all of their life. This imprecision in the semantics is by no means a "fault" restricted to symbiosis. One can find the same multiplicity of meaning in many central terms and expressions in science. Species, heredity, gene, the struggle for existence, the inheritance of acquired characteristics, Darwinism—all have had multiple meanings.

Despite scientists' intentions to invent technical terms that are untainted by the vagaries of ordinary language and cultural contexts, historians, philosophers, and sociologists of science have come to recognize that such ideal language devoid of multiple meanings is, in fact, remote in practice.[6] Change in meaning may occur consciously or unconsciously, accompanying the shifts in paradigms that we associate with scientific revolutions. Moreover, insofar as these conceptual modifications are not monolithic, meanings do not simply change but actually accumulate. Different meanings may also reflect different usages—to describe phenomena in different disciplines.

Because ways of thinking about the natural world and ways of thinking about human social relations, are often inseparable, key terms offer a vantage point for viewing the complex interactions and coupled evolution of science and culture. The permeation of Darwin's theory of evolution by Adam Smith's theory of the market and the division of labor and by Thomas Malthus's law of the relation of population and food supply has been, and continues to be, the subject of detailed analysis.[7] The principle of division of labor has been used to account for almost every level of biological organization, from interspecies relations in ecology to intercellular relations in developmental biology to intracellular relations in cytology and genetics. To see the structure of natural forms as a mirror of the structure of human groups, their inventions, and their intentions has remained imaginatively powerful. Nowhere is this more evident than in the history of symbiosis.

Discussions of anthropomorphism and teleology have been central to debates over the semantics of symbiosis for more than a century. A myriad of metaphors of slavery and consortium, the relations between men and women, among nations, and the relations between humans and domesticated plants and animals informed how one was to understand living arrangements between species. Evolutionary narratives based on these categories were crucial for theorizing on the origin of symbiotic complexes.[8] From the late nineteeth to the late twentieth century such discussions went full circle, and symbiosis was frequently offered as a model for the study of our human social evolution and as a prescription for our relations with the Earth.

The term symbiosis continues to spill over into various domains. One can find it used in studies of the evolution of certain forms of architecture and in cinematography. The origin of "civilization," with the domestication of plants and animals, has also been considered as a natural evolutionary process based on symbiosis.[9] Even "automated theorem proving and logic programming" in computer science has been referred to as a "natural symbiosis."[10] Again, in this context, symbiosis connotes not only naturalization and progress but also the morality of fair play for mutual gain, as in an article in my own field, "Historians and Small Presses: An Emerging Symbiosis?"[11]

For more than a hundred years symbiosis has remained a metadisciplinary discourse; no institutionally defined groups monopolized authority over its meaning or harnessed a practice based on its definition per se—as geneticists did for heredity, for example. In a historical sense, one can think of symbiosis less as some inherent quality or set of qualities than as a number of ways in which biologists relate themselves to it. Those few biologists who converged on the role of symbiosis in evolution came from almost as many backgrounds, and with as many beliefs, interests, and aims, as those who criticized them. Some were mystics, some were mechanists. For some, symbiosis was distinct from and contradicted individual life struggle; for others, it embodied it. For some, it implied a universal mechanism for the synthesis of new individuals and the origin of species; for others, it would always be a supplementary principle. For still others, it represented an exemplar for understanding the totality of the natural world. But the *oneness* some saw in symbiotic associations has to be evaluated in terms of the "*won-ness*" of scientific knowledge.

The history of symbiosis research is a window from which to scrutinize the effects of ever-increasing specialization in the life sciences. It is well recognized, especially by those who have studied scientific controversy, that specialization and discipline formation are not simply the result of a functional division of labor. This can be viewed rather as a strategy for gaining and maintaining adherents to a particular theory or practice and as a locus for funding. Specialization affects the way in which concepts are constructed and nature is classified, ordered, and parceled; it also affects the way in which knowledge claims are assessed. By its very synthetic nature, by transgressing the well-defined boundaries of the individual, symbiosis as a source of evolutionary innovation transgressed disciplinary boundaries and confronted the concepts, doctrines, techniques, and aims of the major biological disciplines—botany, zoology, bacteriology, physiology, immunology, parasitology, cytology, embryology, ecology, genetics, and virology.

A story about the scope of integration in nature and its underlying dynamics as a prescription for our relations with each other, and with other species, the history of symbiosis also offers a critique of the dominant relations which hold between science and nature. It's about constructing concepts of germs, not simply as disease-causing, but as life-giving entities; about understanding microbes as natural historical entities and not simply exploiting them as technologies for biochemistry and genetics; about constructing balanced interdependence as opposed to struggling for independence.

But studies of symbiosis as a source of evolutionary innovation were also carried out against a background of conflict and competition among individual symbiosis researchers. Divided by struggles for priority, geographic distance, language, and war, investigations of the physiological and morphological effects of symbiosis were carried out only on an individual basis for most of the century. Thus unfolds an ironic tale of the fierce individualism of many personalities who pointed to the creative power of associations in evolutionary change.

1

Symbiosis: Evolution in Action

Parasitism, mutualism, lichenism etc., are each special cases of that one general association for which the term symbiosis is proposed as the collective name.

ANTON DE BARY, 1878[1]

The Origins of the Word

Lichenologists of the late nineteenth century frequently told stories of how lichens perhaps more than any other groups of plants had been neglected by previous generations of botanists, how the few who did study them did so on the side and subjected them to great abuse through their careless study.[2] The reasons for this, it was argued, were not hard to find. These plants possessed few qualities to make them conspicuous; as a rule, they are neither striking in color, size, or form, nor do they possess marked useful or harmful properties.

Some lichens had been claimed to be useful in the arts and medicine. During the wars of 1812 and 1815 fevers of all sorts were prevalent in military hospitals. Quinine, which was the popular remedy, became scarce because of the enormous quantities consumed and because of the commercial blockade of Europe. The Austrian government therefore offered a prize of five hundred ducats for the discovery of a cheap, readily available substitute for quinine. The prize was won when lichen was proposed as a substitute, but it was soon found that the medicinal properties of lichen were unsatisfactory and the remedy was abandoned by physicians. Nonetheless, the peasantry of various countries still believe in the healing properties of various lichens. In Sweden *Peltigera aphthosa* was boiled in milk and given to children afflicted with "thrush." Decoctions of various lichens were employed for intermittent fevers. Sir John Franklin and his companions made use of this lichen during their arctic voyages, and its use to prevent the disease known as "Iceland scurvy" continued during the nineteenth century. French and Scandinavian chemists employed lichens in the manufacture of alcohol, but chemical investigations were conducted primarily with a view to improve the dye industry dependent upon the various coloring substances derived from lichens. France, more than any other country, improved upon the methods of extracting the dye, as well as applying it.[3]

During the first half of the nineteenth century, when lichens began to receive some attention from botanical systematists, varied opinions were held with regard to their origin, nature, and position in the vegetable kingdom. When the belief in spontaneous generation was prevalent early in the century, it was often maintained that lichens, as well as algae and mosses, could develop without seed from decomposing water, induced by warmth of sunlight. Others suggested that lichens were mosses that had been checked in their evolution, and the chaotic mass could be considered as "vegetable monstrosities."[4] Lichens had been indiscriminately classified as mosses, algae, or fungi. But by the 1860s, with microscopical aid, lichenologists and cryptogamists in general came to consider them as constituting a distinct class of plants, definitely separated from algae and fungi. Systematists counted 1051 species of lichens in Germany and Switzerland alone.[5] Nonetheless, the origin, nature, and rightful classification of lichens remained a subject of intense controversy.

When, in 1868, the Swiss botanist Simon Schwendener put forth his "dual hypothesis" that all lichens were actually an association of a fungus and an alga, his suggestion was met with bitter opposition.[6] This resistance resulted in part from the kind of social imagery he employed. Lichens, he argued, represented a master–slave relationship. The master was a fungus of the order Ascomycetes, "a parasite which is accustomed to live upon the work of others; its slaves are green algals, which it has sought out or indeed caught hold of, and forced into its service." He went on to describe how the fungus surrounds the alga, "as a spider does its prey, with a fibrous net of narrow meshes, which is gradually converted into an impenetrable covering. While, however, the spider sucks its prey and leaves it lying dead, the fungus incites the algae taken in its web to more rapid activity, nay, to more vigorous increase."[7] Thus the algae kept in slavery are transformed in a few generations to such a point that they are no longer recognizable. Lichens, he argued, had to be considered as modified fungi.

Schwendener's interpretation was an abomination to systematists. The dual nature of lichens threatened not only existing methods of classification but the hard-won autonomy of lichenists themselves. Leading lichenists bitterly and indignantly rejected a view which, as the English Reverend James M. Crombie put it in 1874, pitilessly robbed "their much-loved, if in some respects difficult science," of its well-defined limits which distinguished it from its neighboring classes of cryptogamics.[8]

Throughout the latter part of the nineteenth century many lichenists simply denied the evidence that "deprived lichens of their autonomy" and which had turned them by "the stroke of a magician's wand" into what many, including Crombie, saw as "the unnatural union between a captive Algal damsel and a tyrant Fungal master."[9] Schwendener's theory was "impossible" and absurd, "a useful and invigorating parasitism—who ever before heard of such a thing?"[10] Lichenists continued to classify lichens as a distinct class of plants intermediate between the algae and the fungi.[11]

Despite these protests, research on the dual nature of lichens grew with a new generation of botanists led by Schwendener (1829–1919), who established a leading institute of experimental botany at the University of Berlin, and by Anton de Bary (1831–1888), who did the same at Strassburg.[12] Their studies of lichens were followed by those of several other experimental botanists during the 1870s, 1880s, and 1890s. Most prominent among them were Albert Bernhard Frank, Ernst Stahl, and Johannes Reinke in Germany, Emile Bornet and Gaston Bonnier in France, Andrei Famintsyn in Russia, and Albert Schneider in the United States.[13]

In England, in the 1890s Beatrix Potter also studied the dual nature of lichens as part of her early interest in classifying, dissecting, and drawing fungi. However, unlike the Reverend Crombie, who was able to present his views to the Linnaean Society in 1886,[14] Potter was not. Instead, her work was read before the Linnaean Society of London by her uncle, the distinguished chemist Sir Henry Roscoe, in 1897.[15] Potter could neither deliver the paper herself nor attend the meeting at which it was discussed because women were unwelcome. Her presence as a researcher in the British Museum was also unwelcome, as was the support for the dual nature of lichens she offered to the assistant director of the Royal Botanical Gardens. Following this "storm in a tea kettle," as she called it, she abandoned her interest in a career in science and wrote children's books.

Schwendener's dual hypothesis was confirmed by many botanists who isolated the algae which enter into association with various types of fungi to make particular kinds of lichen.[16] Some even attempted to produce lichens synthetically by culturing the fungus component and the algal constituents, or gonidia as they were called, separately and uniting them to study the formation of the lichen thallus.[17] But isolated cultures of the algae and the fungi were difficult to make. The algae could be cultured separately when supplied with the proper nutrients, but the fungus of the lichen was much more difficult to culture. It was generally reasoned that it had become so closely adapted to the alga that they had more or less lost the ability to live alone. Nonetheless, there were various reported partial successes. Among them, those of de Bary's former student Ernst Stahl[18] at the University of Strassburg and those of Gaston Bonnier[19] at the Laboratory of Plant Biology, Fontainebleau, were the most highly celebrated.

Before this work, it was thought that both the gonidia and the supporting hyphae were the products of development of a single germinating spore. "Gonidia" were supposed to be, as the term indicates, asexual organs of reproduction produced from the hyphae and capable of developing into a new and perfect lichen thallus. (The view that hyphae might also be produced from "gonidia" was often expressed.) The dual hypothesis changed this view radically. The mycelium (hyphae) of the fungus forms the colorless part. The *gonidia* are algal cells essential for the growth of the lichen thallus. The spores of the fungus can begin to germinate alone, but the young thallus very quickly ceases to develop if it does not meet up with the suitable alga. In some lichens the algal and fungal components remained together throughout the life history of the plant. Many lichens propagate by special organs called *soredia,* little globular masses found to be composed of the alga surrounded by some hyphae.

Yet, even many of the experimentalists who accepted and investigated the dual nature of lichens protested against Schwendener's portrayal of the association as a master–slave relationship. Schwendener's interpretation seemed to be belied by his own physiological studies of the fungus and alga in isolation. According to him, the alga brought about synthesis starting from the carbon dioxide of the air; the fungus brought to the alga water and the mineral salts of the soil. The relationship seemed to be much more equitable than Schwendener was prepared to admit. To Johannes Reinke, who subsequently studied various other forms of "cooperative living" involving microorganisms, the relationship of the alga to the fungus was that of the leaves and roots of a green plant. The alga (autotrophic) synthesizes carbohydrates and borrows from the fungus (heterotrophic) the nitrogenous and albuminoid material that the latter builds up with the help of the carbohydrates furnished by the alga; besides this, the fungus draws up water and mineral substances. Reinke suggested the use of the term "consortium" to express the relationship.[20]

What was at stake in these different interpretations was not simply a question of sentimentalism and aesthetics, but the whole question of what constitutes biological autonomy. To Reinke, the fact that the algae could be cultivated artificially did not indicate that lichens could be considered as fungi parasitically associated with algae, as it did to Schwendener. The fact remained that when either of the symbionts is removed the lichen no longer exists; its autonomy is destroyed. Thus Reinke opposed Schwendener's classification of the lichen as a modified fungi. Although double organisms, lichens were morphologically and physiologically sufficiently distinct from both fungi and algae to be recognized as a distinct class. He also pointed to the polyphyletic origin of lichens. The various groups (usually generic in his view) of lichens were derived from different fungal ancestors.[21]

Many of those who studied the dual nature of lichens agreed that the term parasitism with its connotations of disease and destruction did not adequately describe this relationship. Some came to see in the lichen the possibility of a more general phenomenon: associations between phylogenetically distinct organism that ranged from the loosest to the most intimate and essential, and from the most antagonistic and one-sided to the most beneficial for the well-being of both associates. A neutral term was required that did not prejudge such relationships as parasitic. Therefore, in 1877, Albert Bernhard Frank (1839–1900) at Leipzig coined the word *Symbiotismus:* "We must bring all the cases where two different species live on or in one another under a comprehensive concept which does not consider the role which the two individuals play but is based on the mere coexistence and for which the term Symbiosis [*Symbiotismus*] is to be recommended."[22]

Frank carried out detailed studies of various lichens and put forward the suggestion that there might be cases in which the algal component also had become so adapted to the fungus that it could no longer live independently. He subsequently investigated various other kinds of symbioses between microbes and plants and brought them together in his *Lehrbuch der Botanik* (1892), which was frequently cited by those nineteenth-century botanists who concerned themselves with symbiosis.[23] Though Frank's studies were well-known, and he became one of the chief advocates of the view that many associations involving microorganisms could not be labelled parasitism, the origin of the term "symbiosis" was not attributed to him. Instead, it was credited to Anton de Bary.

De Bary was one of the principal researchers responsible for the development of experimental botany in Germany during the 1860s and 1870s.[24] He had set up the first two botanical laboratories at Freiburg and Halle; he was the editor of one of Germany's major botanical journals, *Botanische Zeitung;* and he was also well-known for his classic book *Morphology and Physiology of the Fungi, Lichens and Myxomycetes,* first published in 1866. During the 1870s de Bary established one of the world's leading botanical research centers at the University of Strassburg. It attracted students from England and the United States as well as Germany. Many became influential in botanical studies. De Bary's students Ernst Stahl and Andreas Schimper were well-known for their investigations of plant physiology and symbiosis.

De Bary first used the term symbiosis (*Symbiose*) in an address entitled "The Phenomena of Symbiosis" delivered at a general meeting of the Association of German Naturalists and Physicians at Cassel in 1878. He defined it as "the living together of unlike named organisms."[25] Although he never referred to Frank in this address, in other writings he did refer readers to Frank's article of 1877 in which he had first employed the term. He also supported Frank's view that some species of algae may have become so adapted to lichenism that they could no longer live outside the lichen combination.[26]

Under the rubric of symbiosis de Bary included various kinds of complex associations, ranging along a continuum from parasitic relations to relations in which the associates helped each other. The latter he referred to as "mutualism." The term mutualism had been introduced into biology a few years earlier by the Belgian zoologist Pierre-Joseph van Beneden (1809–1888) at the Catholic University of Louvain, in a lecture to the Royal Academy of Belgium called "A Word on the Social Life of Lower Animals" (1873).[27] This was later elaborated in his popular book *Les Commensaux et les Parasites,* which was translated into German and into English as *Animal Parasites and Messmates* (1876).[28] In it, van Beneden argued against the view, common in his day, that all intimate relations involving "lower animals" living in or on "higher animals" were parasitic. Instead, he argued that the kinds of social relations in animal societies were as varied as those found in human societies, and he classified them in terms of "parasitism," "commensalism," and "mutualism." The parasite, he wrote, "is he whose profession it is to live at the expense of his neighbour, and whose only employment consists in taking advantage of him, but prudently, so as not to endanger his life."[29] The commensal "is he who is received at the table of his neighbour to partake with him of the produce of his day's fishing. . . . The messmate does not live at the expense of his host; all that he desires is a home or his friend's superfluities."[30]

Van Beneden was thinking of various species of fish that swim alongside larger individuals (pilot fish swimming alongside sharks) from which he believed they receive aid and protection. He found many examples of commensals among crustaceans: tiny crabs (pinnotheres) lived inside shells and seemed to exchange food for lodging. Still other "fixed messmates," after choosing a host and installing themselves within it, "completely lose their former appearance: not only do they throw aside their oars and their pincers, but they cease sometimes to keep up any communication with the outer world, and even give up the most precious organs of animal life, not even excepting those of the senses; they are installed for life, and their fate is bound up with the host which gives them shelter."[31] Van Beneden referred to crustaceans named cirripedes, which lodge on the back of a whale or the fin of a shark and subsequently become transformed in such a way that they no longer have a mouth by which to feed and are reduced to a mere case that shelters their progeny.[32] In his view, there existed "almost insensible gradations of differences between parasite, messmate and free animals."[33] The mutualists

> are animals which live on each other, without being either parasites or messmates; many of them are towed along by others; some render each other mutual services, others again take advantage of some assistance which their companions can give them; some afford each other an asylum, and some are found which have sympathetic bonds which always draw them together.[34]

Van Beneden argued that many insects—those sheltering in the fur of mammals or the down of birds as well as crustaceans on fishes—cared for the toilet of their host by feeding on epidermal debris, excretions, infusoria living in the rectum of frogs, and the like. There were other mutualists which rendered services he compared to medical attendance. He mentioned the Egyptian plover, "which keeps the teeth of the crocodile clean,"[35] and he described a certain worm that lives in a lobster and eats only the dead eggs and the embryos, the decomposition of which might be fatal to the host lobster and its progeny.[36]

De Bary cited these definitions and looked for analogous associations among plants. He argued that in the strict sense of van Beneden, one could not have commensalism in the plant kingdom. There were, however, cases approaching van Beneden's mutualists. He described the intimate association of blue-green algae, *Anabaena,* and a genus of small aquatic ferns, called *Azolla,* that float on the surface of freshwater ponds and marshes. Each leaf of azolla is divided into a dorsal and ventral lobe. The ventral lobe floats on the water, and the algae live inside special sealed cavities of the leaves. De Bary noted that no stage of the life cycle of the fern was free of the algae and that it did no harm to the azolla. He suspected that the host protected the algae, but he had "no idea of a reciprocal service they rendered in return."[37] He described a similar case where blue-green algae live in specialized branches of the roots of certain tropical seed plants called cycads. The algae, he pointed out, penetrated into the cells of the roots, causing characteristic changes in their structure and producing nodules in which they lived.[38]

Finally, de Bary came to the case of lichens, which he himself had begun to study more than a decade earlier. In 1866, two years before Schwendener first suggested it, de Bary put forward the claim that at least some (gelatinous) lichens were associations of algae and fungi; he soon recognized this to be true for all lichens. In his address of 1878, he referred to the detailed studies of his former student Ernst Stahl, who, just the year before, had published his results on constructing lichens synthetically.[39] Lichens, de Bary remarked, were "a form of vegetation comprising thousands of species, in which all the individuals not only show an association of two or even three different species, but are constituted only by this association."[40]

But in de Bary's view, not all lichens were comprised of mutualistic associations. One could find a graded series of relations ranging from parasitism to mutualism, according to the association of different species of fungi and algae. This, he suggested, was true for all cases of symbiosis. It was equally difficult to distinguish "the association of symbionts that are tightly united for their common economy from the diverse relations between different organisms which one can discuss under the name of *sociability.*"[41]

When making this remark about "sociability," de Bary recognized that the term symbiosis might equally apply to looser associations such as that between pollinating insects and flowers and those between animals that search for food or shelter and the animals and plants that supply it. He had no objection against making this generalization. To the contrary, he wanted to show that all these phenomena were related.[42] This was a strategic argument that was designed to ensure that lichens were not dismissed as exceptions. Just as parasites would appear to be exceptions if one considered them in isolation of other phenomena of life struggle, he argued, so too would lichens, if they were not considered in relation to other examples of mutual aid.[43]

The most significant aspect of symbiosis, as de Bary conceived of it, was that it could lead to morphological variations that were not pathological. This was as true for the symbiotic algae of cycads and in azolla as it was for the algal and fungal association in lichens. In these cases, there was "no trace of any diminution of vital energy . . . or any other indications of a sick state."[44] These examples, he asserted, were only illustrative of many more cases of morphological variations that could not be explained pathologically. Equally important, "the experimenter could at will prevent the variations or make them appear, by the separation or reunion of the symbionts."[45] Thus he came to a central reason for his lecture—symbiosis was a major source of evolutionary novelty that could and should be investigated experimentally. It was a mechanism of evolutionary innovation in addition to gradual evolution by natural selection based on the accumulation of individual variations within populations of species. However, de Bary was cautious and said nothing about the relative importance of these evolutionary mechanisms: "Whatever importance one wants to attach to natural selection for the gradual transformation of species, it is desirable to see yet another field opening itself up to experimentation. This is why I wanted to call your attention to these here, though they can only shed light on a part of the phenomena."[46]

Making Associations

During the 1880s many botanists followed de Bary and employed the scheme outlined by ven Beneden to classify various functional relationships involving microorganisms. Almost immediately after the publication of de Bary's address, several cases of intimate associations between phylogenetically distinct organisms were brought forward as possible examples of mutualistic symbiosis comparable to lichens. Some of the most widely discussed cases emerged from studies of so-called animal chlorophyll in the simplest forms of life.[47] The main interest in these chlorophyll-containing "animals" lay in their bearing on the long-disputed taxonomic relations between plants and animals.

The early Darwinians, led by Ernst Haeckel (1834–1919), had investigated the simplest organisms with the hope of gaining insight into the early evolution of life on earth. The green inclusions of freshwater "animals," in innumerable shelled and naked amoebae, infusorians, hydras, and turbellarians and in the green branches and crusts of freshwater sponges, were long familiar phenomena. Studies by Haeckel, Johannes Müller, Louis Agassiz, and T. H. Huxley had also shown that luminous yellow to brown formations occurred regularly in all kinds of marine animals: protozoans, sponges, coelenterates, and turbellarians. Immediately, the green of plants had come to mind, and during the middle of the century, it became clear that the green color of various animals was indeed chlorophyll.

The major dichotomy of plants and animals did not seem to hold true for these chlorophyll-containing organisms. Some possessed locomotion like animals yet had modes of living that were more plantlike than animal. In some cases, even one and the same species could nourish itself at one time as a plant, at another as an animal, according to its circumstances. Thus it was supposed that here in the realm of the very small and relatively simple, one might be observing examples of the not-quite-animal–not-quite-plant ancestors of all living things. If this were the case, one would be dealing with a third kingdom. Within a decade, five researchers on both sides of the Atlantic challenged the age-old plant–animal dichotomy by proposing another kingdom. Richard Owen (1859) called it the Protozoa, John Hogg (1860) designated it the Protoctista, Thomas B. Wilson and John Cassin (1863) designated it the Primilia, and Ernst Haeckel (1866) designated it the Protista. "Protista" is the plural superlative form of the Greek word *protos,* "first." To Haeckel protists were the very first living creatures.[48]

The most common belief about the origin of the chlorophyll observed in many different species of invertebrates—from protozoa to planarian worms and sea anemones—was that it was an animal-specific product endogenous in its nature, "animal chlorophyll." During the 1880s, however, this interpretation was challenged. Most of the green freshwater animals could also be found in a colorless state, but they did not turn green when exposed to light, as might be expected if their coloring matter were a product of their own metabolism. Furthermore, these chlorophyll granules, unlike "true chlorophyll bodies" of plants, could be extracted from the animal and yet continue to live. Based on these arguments a number of biologists rejected the hypothesis of "animal chlorophyll" and asserted that all these inclusions were examples of cooperative living of plant symbionts. Some of the most influential evidence in support of this new theory was provided independently by Karl Brandt in Berlin and Patrick Geddes at the University of Edinburgh.

In November 1881, Brandt read a paper giving the results of a series of observations on the *Zusammenleben* of algae and animals before the Physiological Society of Berlin.[49] For the algae found in the bodies of *Hydra, Spongilla, Stentor,* and the like, he created the generic name *Zoochlorella.* For those in the "yellow cells" which occurred in the radiolarians, sea anemones, and so on, he created the generic name *Zooxanthella.* Brandt demonstrated that these algae were capable of carrying on an independent existence after being removed from the animal, and that they were able to produce starch grains. He examined the physiological function of the algae and found them to be of service to the host in supplying food. So long as the animals contain few or no green or yellow algae they are nourished, like true animals, by the absorption of solid organic substances; but as soon as they contain a sufficient quantity of these algae, they are nourished, like true plants, by assimilation of inorganic substances. In the latter case, the algae living in the animals perform all the functions of the chlorophyll bodies of plants. Finally, he compared the modes of life of these compound plant-animals, Phytozoa, with that of lichens. As he saw it, the Phytozoa represented a case of mutual exploitation: morphologically the alga was the parasite, and physiologically the animal was the parasite.

Geddes advanced his theory of "reciprocal accommodation" in a paper entitled "Symbiosis of Alga and Animals" read before the Medical Faculty of the University of Edinburgh as trustees for the quinquennial Ellis Physiology Prize in October 1881. The paper was published in *Nature* the following year in a call for further experimental investigation of similar cases.[50] His own experiments of algae in sea anemones and radiolarians showed him that these algae were not parasites. He related his work to the case of lichens and interpreted them in terms of a theory of mutual interdependence. As he saw it, one could not imagine a more ideal existence for a vegetable cell "than that within the body of an animal cell of sufficient active vitality to manure it with carbonic acid and nitrogen waste, yet of sufficient transparency to allow the free entrance of the necessary light." Conversely, for an animal cell there could be "no more ideal existence than to contain a vegetable cell, constantly removing its waste products, supplying it with oxygen and starch, and being digestible after death." In short, this was "the relation of the animal and vegetable world reduced to the simplest and closest conceivable form."[51]

It was clear to Geddes that the incorporation of algae in animals was an evolutionary adaptation that would favor those animals which possessed them. It could by no means be termed a case of parasitism, for if this were true, "the animals so infested would be weakened, whereas their exceptional success in the struggle for existence is evident." He asserted that *Anthea cereus,* which contains algae, probably outnumbered all the other species of sea anemones put together, and the radiolarians which contain yellow cells were far more abundant than those which lacked them. The nearest analogue to this "partnership" could be found in the vegetable kingdom, "where, as the researches of Schwendener, Bornet, and Stahl have shown, we have certain algae and fungi associating themselves into the colonies we are accustomed to call lichens, so that we may not unfairly call our apicultural Radiolarians and anemones *animal lichens.*" However, in Geddes's view, this association was far more complex than that of the fungus and alga in the lichens; it stood "unique in physiology as the highest development . . . of reciprocity between the animal and vegetable kingdom."[52]

The publication of the papers of Geddes and Brandt led immediately to a priority dispute involving several biologists. The same year Brandt's paper was published, a German translation of an 1876 paper by Hungarian biologist Géza Entz of Klausenburg appeared, detailing various experiments which demonstrated algae living inside infusoria.[53] His paper published in Hungarian was not known and remained inaccessible to the scientific world outside Hungary. However, Geddes had noted an even earlier report: in 1871 Cienkowski had argued that the yellow cells in some radiolarians were algae living within an animal. Entz took second place.[54] Despite this priority dispute, the view that the chlorophyll bodies of these organisms were actually symbiotic algae remained hotly contested throughout the latter part of the nineteenth century.[55] At the same time, the results of Cienkowski, Entz, Brandt, and Geddes were confirmed and extended by others, including some who, like Famintsyn,[56] had investigated lichen symbiosis.

These apparent examples of cooperative living were soon allied with another striking case: microorganisms living in the root tubercles of Leguminosae. The tubercles on clover and the benefit the legumes exerted on soil had been known since ancient times, and Justus Liebig argued that the benefit was based on an increase of nitrogen. By the end of the 1880s, tubercles were found on nearly all legumes, and it had been demonstrated that this was due to soil bacteria which extract nitrogen from the atmosphere.[57] In further studies by Pierre Mazé at the Institut Pasteur, the bacteria were isolated and the nodules were produced synthetically by culturing first the bacteria and second the plant in sterilized soil and seeding the former into the soil. It was clear that the bacteria did the work of fixing nitrogen and the plant took advantage of it; the precise manner in which this worked became the subject of intense investigation.[58]

The root tubercle of legumes, in turn, was soon regarded as functionally analogous to another widespread case involving a root fungus. The presence of fungi on the roots of trees had been known since the middle of the century, but they had been commonly supposed to be of a parasitic nature. However,

A.B. Frank put forward the hypothesis that this relationship was one of mutual assistance. In 1885 he coined the word "mycorrhiza" for the fungus root which he claimed constituted a "Pilzsymbiosis" or "Wurzelsymbiosis."[59] Frank began these studies at the Plant Physiology Institute of the Royal Agriculture Academy in Berlin. They were carried out at the instigation of the German State Forestry Department in connection with a scheme for the development of truffle culture in Prussia.[60] Frank did not advance truffle culture, but he did investigate the relationship of the fungi which he found regularly infecting and frequently distorting the roots of forest trees, including oaks, beech, and conifers. Besides showing the great extent and regular occurrence of mycorrhiza, he distinguished between what he called "ectotrophic" mycorrhizal fungi, which remain external to roots, around which they form a mycelial sleeve, and "endotrophic" mycorrhizal fungi, which penetrate into the cells of the root.

Far from being parasites, Frank suggested that plants actually attracted the fungi by what he considered to be some secretion. Just as insectivore plants captured insects, "fungivore plants" captured fungi to procure nitrogen. He made a similar interpretation for nitrogen-fixing bacteria and legumes. In the case of ectrophic mycorrhizal fungi and plants bearing them, he suggested that the fungus was a functional substitute for the root hairs: it must imbibe from the soil and bring to the plant both mineral salts and organic nitrogenous food from the humus: the plant, for its part, must yield carbohydrates, which it has built up, to the fungus. The endotrophic fungus would, in addition, make a final contribution to the nutrition of the plant in being digested by it and thus providing it with nitrogen. Thus it seemed to him that endotrophic fungi benefited trees without receiving anything in return.

The mycorrhiza that attracted most attention during the late nineteenth century was that of orchids.[61] By the 1890s fungi in the cortical cells of the roots of orchids had been identified in about five hundred species of orchids. Research on the nature of the association culminated around the turn of the century with the celebrated work of the French botanist Noël Bernard, who was credited with showing their importance and their exact role in the life of the plant.[62] Control over the germination of orchids had been unsuccessful during the nineteenth century, that is, except by techniques that were kept secret because of their commercial value. Bernard isolated mycorrhizal fungi in pure culture and demonstrated that the penetration of the fungi made it possible for the seed of the orchid to germinate. He compared the action of fungi on orchid embryos to the action of spermatozoa on eggs.[63]

Bernard's work could easily account for the success and failures of previous workers. It had been known that in order to get the seed to germinate it was necessary to sow it in the soil of a pot in which the parent plant had been growing, that is, in the soil containing the fungus. The germination of orchids in greenhouses became easier, because the soil there had gradually become richer in fungi as a result of prolonged culture. In natural conditions the immense number of seeds compensated as far as the perpetuation of the species was concerned for the loss of numerous embryos which did not meet the fungus necessary for their development. Bernard also showed that orchid seeds would germinate on sterile sugar-containing media, a discovery that later provided the basis of the commercial techniques for raising orchids from seed.[64] But it was the social relations between such organisms, not the immediate practical or commercial significance of symbiosis, that attracted most attention from biologists.

By the end of the 1880s, those who had studied intimate symbiotic associations believed that they could detect, as de Bary put it, "every conceivable gradation . . . between the parasitism which quickly destroys its victim and that in which parasite and host mutually and permanently further and support one another,—the relation which is most conspicuous in the formation of Lichens and which Van Beneden has termed *mutualism.*"[65] Mutualism itself was developed from diverse perspectives. In order to better understand the ways in which nineteenth-century biologists understood and discussed its meaning we first need to place this discourse in the larger context of evolutionary debates surrounding "the struggle for existence" and "man's place in nature."

2

The Meanings of Mutualism

> Ethically, there is nothing in the phenomena of symbiosis to justify the sentimentalism they have excited in certain writers. Practically, in some instances, symbiosis seems to result in mutual advantage. In all cases it results advantageously to one of the parties, and we can never be sure that the other would not have been nearly as well off, if left to itself.
>
> ROSCOE POUND, 1893[1]

During the late nineteenth century struggle and competition were elaborated equally in the "natural" and the human social realms. The use of natural law as the basis for a given view of society became commonplace in social, political, and economic theory.[2] Studies of symbiosis, insofar as they implied something other than conflict, became identified with mutualism. There was no coherent theory for the origin and nature of mutualistic associations. Some emphasized cooperation within species; others emphasized cooperation between them. Collectively, they were considered to oppose the main focus of Darwinian evolutionists on illustrating conflict and competition. The result was a transformation in the meaning of the term symbiosis. A restricted definition of symbiosis as mutualism emerged in opposition to a restricted view of "the struggle for existence" in terms of nature as "red in tooth and claw."

The Political Context of Mutualism

In *On the Origin of Species* (1859), Darwin premised that he used the expression "Struggle for Existence in a large and metaphorical sense, including dependence of one being on another, and including (which is more important) not only the life of the individual, but success in leaving progeny."[3] He discussed the mutual benefits of flowers and the insects which pollinate them when collecting nectar[4] and he mentioned the dispersal of mistletoe seeds by birds.[5] Nonetheless, his fundamental ecological and evolutionary thinking and that of those who followed him was that adaptation and speciation could be explained by conflict and competition.

In 1838, Darwin had read *Essay on the Principle of Population,* in which Thomas Malthus asserted that populations, if left unchecked, grew geometrically while food supply grew arithmetically. And in *The Origin,* Darwin affiliated his theory of natural selection with Malthus's *Essay* to argue for the intensity with which problems of subsistence press upon species and for the importance of the struggle for existence for evolutionary change.[6] Darwin argued that the existence of heritable variations within a species, together with the production of more offspring than could possibly survive, constituted the conditions under which "favourable variations" would be preserved and "injurious variations" destroyed. This "natural selection" was the basic cause of adaptive change.

Malthus was a cleric and professor of history and political economy at Haileybury College. His ideas had two principal determinants. In the first place his *Essay* of 1798 was a reaction against the optimistic ideals of the Enlightenment and the utopian views of such philosophers as Condorcet and Godwin, who contemplated indefinite progress toward the complete absence of struggle among men, no illness, no sexual urge, no cares. Malthus argued that it would never be possible to realize such ideals, for they took no account of an absolutely fundamental issue: the problem of population growth. Maintaining that there would always be too many mouths for the world to feed, Malthus reasoned that mankind must always be subject to famine, poverty, disease, and war unless some means of limiting population could be found. As Robert Young has argued: "Godwin had gone too far in removing man from nature. Malthus's reaction provided the essential change of perspective for putting man into nature once and for all."[7]

Malthus's *Essay* was also a reaction against restructuring of the poor laws in England. His attack on the poor laws and criticism of public charity became more prominent in the more widely read second edition of his *Essay* of 1803. He maintained that giving people extra financial support would encourage them to breed more, thus exacerbating the poverty problem. For Malthus and his followers state charity would mean the end to individual industry. If there were more charity, the demand would rise to exhaust it. Everyone would eventually be dragged down together.

Darwin's theory of natural selection can by no means be reduced simply to Malthusian principles. Nonetheless, in *The Origin* he characterized the struggle for existence as "the doctrine of Malthus applied with manifold force to the whole animal and vegetable kingdoms."[8] The struggle for existence in its strict sense applied to relations between species as well as within them. "But, the struggle almost invariably will be most severe between individuals of the same species, for they frequent the same districts, require the same food, and are exposed to the same dangers."[9] Competition led to suffering, death, and extinction because there were always too many mouths for the world to feed. But this suffering, by producing "fitter" individuals through natural selection, would ultimately produce better organisms and lead to evolutionary progress. Competition was both inevitable and desirable. As Darwin concluded in *The Origin.* "Thus from the war of nature, from famine and death, the most exalted object which we are capable of conceiving, namely, the production of the higher animals, directly follows."[10]

While competition and progress through individual life struggle were dominant themes of both natural and social science in the nineteenth century, political and intellectual opposition developed in concert. In Britain, various associations such as trade unions, Chartist groups, and the "Friendly Societies" were formed to allow workers to deal with catastrophes such as illness or funerals. The analogous organizations in France were the mutual aid associations. Beginning in the period of the Revolution, the different *Mutualité* societies were a hotbed of socialist ideas. French *mutuellisme*'s most celebrated exponent was Pierre-Joseph Proudhon, who became famous for his book of 1840, *What Is Property?* To which he answered simply "property is theft." By this, he meant only property in its Roman law sense of right of "use and abuse." In property rights he saw the best protection against the encroachment of the state. Proudhon is recognized today as one of the founders of socialist and anarchist movements. Along with Karl Marx, Robert Owen, and others, he repudiated Malthus for laying at the door of the poor the responsibility for their own plight, which on the contrary was the result of the selfishness of the rich.[11]

Proudhon's mutualism was an antiauthoritarian ideology based on the abolition of governments and the reconstruction of society as an overarching federation of workers' cooperatives. He denounced political revolution as unnecessary and even dangerous to liberty. He believed that the path to socialism could be blazed through the development of a system of mutual credit, through which workers could borrow the funds to amass capital and create cooperatives, which would eventually replace capitalism. His aim was to render capital incapable of earning interest, by establishing a national bank based on mutual confidence of all those engaged in production, who would agree to exchange their products at cost value based on the number of hours of labor required to produce a given commodity. Under this system, which Proudhon described as *mutuellisme,* all the exchanges of services would be strictly equivalent.

Regarded by the French government as an exceedingly dangerous man, Proudhon was imprisoned from 1849 to 1852 for publishing articles criticizing Louis Napoleon. From 1858 to 1862 he lived in exile in Belgium. Pardoned by Napoleon III, he returned to Paris in 1862 and continued to gain influence among the Paris workers with his mutualist ideas. In short, the Proudhonians' idea of socialism was one of just and equal exchange, made possible by eliminating the unfair advantage gained by the capitalists through their inheritance of property. Mutual credit would allow workers to produce on equal terms with capitalists, since the accumulated wealth would disappear. Credit would be freely available to all.

In 1871 the Commune took control of Paris, and in the brief period it had to reorganize the economy before being bloodily repressed, it manifested a clear Proudhonian viewpoint. The metalworkers and mechanics unions of Paris expressed their aims: "The abolition of the exploitation of man by man, last vestiges of slavery; The organization of labour in mutual associations with collective and inalienable capital."[12] The Association of Women, led by the revolutionary socialist Elizabeth Dimitrieff, proposed organizations that would help provide work for women and would "instil into them a strong consciousness of mutualism."[13]

This was the sociopolitical background against which Pierre-Joseph van Beneden (1809–1893) introduced the term mutualism. Van Beneden drew analogies from industry, human social relations, and morality to describe the social relations he saw in nature. For example, when making analogies between the "lower animals" and human societies, he compared industrialists leading the life of noblemen to parasites:

> In the ancient as well as the new world, more than one animal resembles somewhat the sharper leading the life of a great nobleman; and it is not rare to find, by the side of the humble pickpocket, the audacious brigand of the high road, who lives solely on blood and carnage. A great proportion of these creatures always escape, either by cunning, by audacity, or by superior villainy, from social retribution.[14]

The word "sharper" is used as the English translation for *chevalier d'industrie* in the French version. The implication in both is that the industrialist is a thief, and Douglas Boucher has remarked that "the ironic comparison of parasites and 'knights of industry' is tantalizingly close to 'property is theft.' "[15] Moreover, Boucher suggests that the use of the term "mutualists" itself may well have evoked thoughts of Proudhon and the mutualists of the Commune repressed only a few years earlier.[16]

The Plover and the Crocodile

Van Beneden was a Catholic with deep religious convictions.[17] His mutualisms were imbued with ideas from natural theology; they were examples of perfect adaptations created by the divine wisdom of God:

> All these mutual adaptations are pre-arranged, and as far as we are concerned, we cannot divest ourselves of the idea that the earth has been prepared successively for plants, animals, and man. When God first elaborated matter, He had evidently that being in view who was intended at some future day to raise his thoughts to Him, and do Him homage.[18]

Mutualisms were frequently used by ancient writers as examples of Nature's balance: those tendencies which prevented any species from becoming too abundant or going extinct were due to divine providence. Herodotus told the story about a mutually beneficial relationship between Nile crocodiles and a species of plover. The plover ate leeches from the crocodiles' mouth; the crocodile never hurt the bird.[19] Aristotle liked that story (as did van Beneden) and mentioned it in three different treatises, and he had also reported (as did van Beneden) that a mutual relationship existed between the bivalve pinna and the crustacean pinnotheres.[20] Similar descriptions were given by Cicero and Aelian, who drew the moral that humans should learn friendship from nature.[21] Pliny also told that "friendships occur between peacocks and pigeons, turtle-doves and parrots, blackbirds and turtle-doves, the crow and the little heron in a joint enmity against the fox kind, and the goshawk and kite against the buzzard."[22]

Mutual interactions were favorite examples of Divine Providence in the natural theology of the seventeenth and eighteenth centuries which found its ultimate expression in the works of Linnaeus. By the economy of nature, Linnaeus wrote, "we understand the all-wise disposition of the Creator in relation to natural things, by which they are fitted to produce general ends, and reciprocal uses."[23] In this arrangement, living beings were so connected, so chained together, that they all aimed at the same goal. The search for "general ends," for an overriding purpose and agency in nature, was the crucial impetus to the Linnaean school of natural history. Besides Linnaeus's pivotal essay of 1749, "The Oeconomy of Nature," the leading works in this field included John Ray's *The Wisdom of God Manifested in the Works of Creation* (1691), William Derham's *Physico-Theology* (1713), and William Paley's *Natural Theology* (1802) and the *Bridgewater Treatises* (1833–36). All were united by their common repudiation of doctrinal schisms and of claims of private mystical revelations, and instead looked to reason and the testimony of nature to establish their faith on a firm, universally acceptable ground.[24]

Followers of natural theology opposed the views of the seventeenth-century philosopher Thomas Hobbes, who, in his famous political treatise *The Leviathan* (1651), represented the state of nature as a war of all against all. Hobbes argued that without a powerful government—an almighty Leviathan to restrain human life—men would live as animals without virtue and morality, without agriculture, arts, and letters. Donald Worster emphasized that the Linnaeans implicitly accepted the assumptions underlying the Hobbesian view but argued that Nature did in fact have laws which prevented such disorder and disharmony.[25] The Creator, they maintained, had established a vast system of subordination to ensure peace in the natural world. Each species had been assigned a fixed place in a social hierarchy or scale of being. The chain of being was a system of economic interdependence and mutual assistance. Even the most exalted creatures must depend upon those lower on the scale for their very existence; man and worm alike lived to preserve each other's life.

All of these views can be found explicit in van Beneden's *Animal Parasites and Messmates*. "The assistance rendered by animals to each other," van Beneden argued, "is as varied as that which is found among men. Some receive merely an abode, others nourishment, others again food and shelter; we find a perfect system of board and lodging combined with philozoic institutions arranged in the most perfect manner."[26] Within this system of mutual aid some organisms carried out medical attendance, others carried out roles of menagerie keepers, some cleansed the animals themselves; others kept their cages clean and removed the dung and filth. Paupers helped other paupers as well as the "higher classes." Some members of the higher classes helped the lower, as pinnotheres helped mussels in which they took shelter by dropping crumbs of food from their pincers. They were not all like the "rich man who installs himself in the dwelling of the poor, and causes him to participate in all the advantages of his position." The pinnotheres were good lodgers. But the point was equally true that they, in turn, required assistance from the lower classes: the "noble crab" relied on lodging from the lower "blind and legless mussel."[27]

Van Beneden remained a lifelong opponent of the view that evolution resulted from a "struggle for life" and natural selection. If this were true, he wondered, how is it that those beings which were powerfully armed for struggle, the giants of the animal world of diverse classes, were precisely those which had succumbed in the struggle for existence?[28] The idea that an individual struggle for life could lead to extinctions but not to evolutionary progress was not uncommon among those who studied mutualistic symbiosis. Van Beneden's definitions were taken up by zoologists as well as botanists. Alfred Espinas used them in his doctoral dissertation presented at the University of Paris in 1877; his book *On Animal Society,*[29] translated into English in 1935, discussed tick-birds and rhinoceroses, ants and aphids, mixed flocks of birds, and especially domesticated animals, as examples of mutualisms.

Although many botanists and zoologists came to use the terminology borrowed from van Beneden, the meaning they placed on it generally differed. For de Bary, and many others, mutualistic and parasitic associations involved no finality. They were not categories laid down by Divine Providence. Nonetheless, mutualism was still posited either explicitly or implicitly in direct opposition to a Hobbesian war of each against all and a belief in evolutionary and social progress resulting from a pitiless struggle for individual advantage. These views were expressed most prominently in the writings of the Russian anarchist Peter Kropotkin (1842–1921), which culminated in his best-selling book, *Mutual Aid: A Factor of Evolution.*

Protesting the Gladiator's Show

Kropotkin grew up in the midst of the revolutionary movement against the Russian tsars, in the years of intense struggle for the abolition of serfdom and the establishment of a constitutional government. He was born a prince of the old nobility of Moscow, and at twenty he became an officer in the army. He studied mathematics and geography at the University of St. Petersburg for five years. At age thirty he became secretary of the section of the Russian Geographical Society dealing with physical geography, but he refused the appointment of secretary of the whole society because he was drawn to the cause of peasants. The discovery that he was engaged in revolutionary activities while at university caused a sensation. He was arrested and held in prison without trial. After a year's confinement, he escaped and found refuge in England. He remained in exile forty-two years, engaged in scientific research and anarchist propagandizing.[30]

In England, Kropotkin was well recognized for his work in geography. He was elected to the British Royal Geographical Society, an honor which he declined because of his hostility to any association with a "royal" affiliation. He was offered the chair of geography at Cambridge but declined the offer since it was plain that the university would expect him to cease his anarchist activities while in its service. Instead, he earned a living through his scientific writings.[31]

Mutual Aid, Kropotkin's most popular work, a classic reply to the school of the "survival of the fittest," was an attempt to make the case for voluntary cooperation and freedom on a scientific basis and to embed anarchism in evolutionary theory. One chapter is devoted to its influence in animal societies, the rest of the book to a historical study of its growth and power from "primitive" tribes to the present. First published in 1902, *Mutual Aid* was based on a series of articles written, beginning in 1890, for the widely read magazine *The Nineteenth Century.* They were a response to an article published by one of England's leading evolutionary authorities, "Darwin's bulldog," Thomas Huxley. In 1888, Huxley issued what Kropotkin called his "Struggle for Life Manifesto": "Struggle for Existence and Its Bearing upon Man." From the point of view of the moralist, Huxley wrote,

> the animal world is on about the same level as the gladiator's show. The creatures are fairly well-treated, and set to fight—whereby the strongest, the swiftest, and the cunningest live to fight another day. The spectator has no need to turn his thumbs down, as no quarter is given.

Huxley continued that among animals so among men,

> the weakest and stupidest went to the wall, while the toughest and shrewdest, those who were best fitted to cope with their circumstances survived. Life was a free fight, and beyond the limited and temporary relations of the family, the Hobbesian war of each against all was the normal state of existence.[32]

In his famous Romanes Lecture on "Evolution and Ethics" (1893), Huxley declared that human morality did not have the sanction of nature.[33] Maintaining that there was no trace of moral purpose in nature, he asserted that no ethics were required merely to survive, that the "fittest" to survive in the struggle for existence may be and often are the ethically worst. Human social progress then required a checking of the cosmic process.

Kropotkin protested against the distinction between the nonmoral natural and the moral human social realms. If humans were inherently cooperative, as he assumed them to be, then cooperative behavior and altruistic feelings themselves were important progressive elements in organic evolution. But more than assumptions about human nature underlay his perspectives. They were supported by empirical observations and by a tradition of Russian biologists who rejected Malthusian assumptions underlying Darwinian evolution. In natural history expeditions made in his youth to eastern Siberia and northern Manchuria, Kropotkin had failed to find that bitter struggle for the means of existence among animals of the same species—which Darwinists considered the dominant characteristic of struggle for life and the main factor of evolution. Instead, he saw "Mutual Aid and Mutual Support carried on to an extent" that made him "suspect in it a feature of the greatest importance for the maintenance of life, the preservation of each species, and its further evolution."[34]

Kropotkin was first alerted to the view of mutual aid as a factor in evolution in January 1880, when attending a lecture "On the Law of Mutual Aid" delivered at a Russian Congress of Naturalists by the zoologist K. F. Kessler, then the dean of St. Petersburg University.[35] Daniel Todes argues that Kropotkin's views on mutual aid were actually representative of many evolutionists in Russia. Objections to the dog-eat-dog character of British industrial competition, and to Malthusian principles, could be found on both ends of the Russian political spectrum. But Todes also highlights Kropotkin's own reasoning based on Russia's land and natural history. In an immense underpopulated country, for the most part a harsh land, competition was more likely to find organism pitted against environment than organism against organism. Malthusian principles seemed to be simply irrelevant. This combination of anti-Malthusian and non-Malthusian influences, Todes argues, predisposed Russians against accepting overpopulation and intraspecific conflict as important factors in evolution.[36]

In *Mutual Aid,* Kropotkin asserted that both the meaning and extent of the struggle for existence in evolution had been exaggerated (much to the regret of Darwin himself) while the importance of sociability and social instincts in animals for the well-being of the species and community had been underrated. Yet as he recognized, Darwin himself had used the term "struggle for existence" mainly in its narrow sense. The concept of struggle had become narrower still, Kropotkin remarked, when followers of Darwin "raised the 'pitiless' struggle for personal advantage to the height of a biological principle which man must submit to as well."[37] "Sociability," he argued, was as much a law of nature as was mutual struggle. As he saw it, those animals that practiced mutual aid were much more "fit," intelligent, and highly developed than those that were constantly at war with each other. Thus Kropotkin relayed the message of nature back to human social relations:

> "Don't compete!—competition is always injurious to the species, and you have plenty of resources to avoid it!" That is the *tendency* of nature, not always realised in full, but always present. That is the watchword which comes to us from the bush, the forest, the river, the ocean. "Therefore combine—practice mutual aid! That is the surest means of giving to each and to all the greatest safety, the best guarantee of existence and progress, bodily, intellectual, and moral."[38]

Kropotkin's discussions of mutualism, like those of many biologists in Russia, were concerned primarily with cooperation within animal species rather than between them: herding instincts of deer against a common foe, the colonies built by termites and ants. As he put it, "The ants and the termites have renounced the 'Hobbesian war' and they are the better for it."[39] In fact, Kropotkin actually juxtaposed cooperation among animals belonging to the same species or society to "the immense amount of warfare and extermination going on admidst various species."[40] He made only the briefest mention of microbial cooperation, that is, among "the lowest animals," when he commented that "we must be prepared to learn someday, from the students of microscopic pond life, facts of unconscious mutual support, even from the life of microorganisms."[41]

Symbiosis as Mutualism

Although Kropotkin never mentioned the term "symbiosis" in his *Mutual Aid,* his opposition to the Hobbesian-Malthusian view of the struggle for existence illustrates well the ethical context in which cases of symbiosis were discussed and understood by biologists of the late nineteenth century. Although cooperation between species was frequently not even mentioned by mutual aid theorists who focused on the social life of animals, when 'symbiosis' was discussed it was identified in terms of mutual aid. In Tode's book on Russian evolutionary thought, *Darwin without Malthus,* the word symbiosis is mentioned only twice. But when it was used by the late nineteenth-century biologists he investigated, it was posited in direct opposition to an all-exclusive individual life struggle. Thus Alexander Fedorovich Brandt, director of the Zoological Museum of the Academy of Science, wrote in his essay "Symbiosis and Mutual Aid" (1896) that it was the zoologists' responsibility to show "how the struggle for existence in the animal world was exaggerated and to present, in opposition to it, the principle of mutual aid."[42] Similarly, Mikhail Mikhailovich Filippov wrote in 1894 that "beginning with the phenomenon of so-called symbiosis . . . and ending with the complex conditions of . . . the lives of so-called social animals, we see an entire series of [cooperative] interactions among individuals."[43]

In Britain, those who opposed Huxley's "gladiator's show" frequently advocated a process of progressive evolution toward cooperation among organisms. Some took the organism itself as a model or metaphor for interorganismic integration and cooperation. In ethical terms, this organicist view often implied a repudiation of individual advantage in favor of family and community interests. But the meaning and mechanisms used to account for cooperation frequently differed.

In *The Evolution of Sex* (1889), Patrick Geddes and J. Arthur Thomson argued against laissez-faire individualism upon which they asserted Darwin and his followers took his stand and which ignored the well-being of the individual in considering the advancement of the species.[44] One could not regard competition and the survival of the fittest as the essential mechanism of progress as economists and biologists had both been doing. As they saw it, "the ideals of ethical progress, through love and sociality, co-operation and sacrifice, were the highest expressions of the central evolutionary process of the natural world."[45] They traced the "twin streams of egoism and altruism" to a common origin "in the hunger and reproductive attractions of the simplest forms of life."[46]

Their central thesis was that the general progress of both the plant and animal worlds, and notably the great uplifts, must be viewed not simply in terms of individualism "but very largely in terms of sex and parenthood, of family and association; and hence of gregarious flocks and herds, cooperative packs, evolving tribes, and ultimately of civilized societies." They emphasized that unconscious mutual aid could be found among the simplest forms of life; that "the greatest step in organic nature, that between the single-celled animals and many celled animals . . . is not due to selection of the more individuated, but to the union of the cells into an aggregate whereby each becomes diminishingly competitive and increasingly subordinated to the social whole."[47] The colonial or multicellular forms may have justified "their existence in the struggle for existence, just as unions of many kinds do in human society." But they could neither be accused of any "provision of future advantage in remaining clubbed together in cooperation" nor "be credited with much altruism in doing so." This "greatest of morphological steps" was due to "a process not interpretable in terms of individual advantage."[48] "Each of the greater steps of progress," they argued, "is in fact associated with an increased measure of subordination of individual competition to reproductive or social ends, and of interspecific competition to co-operative associations."[49]

Geddes and Thomson continued to emphasize cooperation (mainly family and species interests) in opposition to individualism in their little book *Evolution* (1911). Again they lamented that "Huxley's tragic vision of 'nature as a gladiatorial show,' and consequently of ethical life and progress as merely superimposed by man, as therefore an interference with the normal order of Nature, is still far too dominant among us."[50] They considered the writings of Kropotkin to be an elaboration of the concluding thesis of their own *Evolution of Sex;* they held the same to be true for the well-known writings of the Scottish preacher Henry Drummond.[51]

Like Kropotkin and Geddes and Thomson, Drummond sought a natural foundation for moral behavior. But his main interest was to save spiritualism from mechanistic interpretations of evolution, which for him was the result of a "providential plan." Drummond had already gained a considerable following with his philosophical book, *Natural Law in the Spiritual World* (1883).[52] In his famous Lowell Lectures, published as *The Ascent of Man* (1894), he emphasized cooperation within species as well as the nutritional bonds between plants and animals to assert that the struggle for life was subordinate to the struggle for the life of others. Drummond's natural history was a love story: "The path of progress and the path of Altruism are one. Evolution is nothing but the evolution of Love, the revelation of Infinite Spirit, the Eternal Life returning to Itself."[53]

In Drummond's scheme, "the struggle for the life of others" in humans began with the family—the basis of human sympathy and solidarity. The formation of groups and societies for mutual protection and mutual help and cooperation had played an important part in evolution. But the cooperation brought about by reproduction, he asserted, was more universal, radical, and efficient.[54] Nearly all the foods of the world were "love foods": the dates, raisins, bananas, honey, eggs, grains, seed cereals; and all the drinks of the world were "love drinks": the juices of the sprouting grain, the withering hop, the milk from the cow, the wine from the grape. "Remember that the Family, the crown of all higher life, is the creation of Love; that cooperation, which means power, which means wealth, which means leisure, which therefore means art and culture, recreation and education, is the gift of Love."[55] All these were the fruits of the altruistic principle of reproduction. There could be no progress without it. All of the natural order was embedded in "otherism."

For Drummond cooperation was largely but not exclusively the "gift of reproduction"; it also resulted from the division of labor. He asserted that "self-sufficiency leads to nothing in evolution."[56] All of progressive evolution from the emergence of the unicellular organisms to multicellular organisms resulted from cooperation. Everything came into being because of something else. There would be no animals without the cooperation of chlorophyll-containing plants "to break up the mineral kingdom and utilize the products as food."[57] The evolution of species, he asserted, was identical to the evolution of the "individual" where "division of labor in new directions arises for the common good; leaves are organized for nutrition, and special cells for reproduction." "A flower is organized for cooperation. It is not an individual entity, but a commune, a most complex social system."[58]

Drummond allied the views of Kropotkin with those of Herbert Spencer to assert that "evolution is primarily the formation of an aggregate."[59] Drummond made no mention of lichens or algae living inside "lower animals" when discussing the integration of the natural world in the *Ascent of Man*. However, Spencer did in the revised edition of his *Principles of Biology* (1899), when arguing that nature as a whole represented a superorganism.

Organicism, Symbiosis, and the Division of Labor

Herbert Spencer (1820–1903), who used evolutionary theory to champion individualism and social policies of laissez-faire, and who coined the expression "survival of the fittest," has often been portrayed by historians as a brutal Social Darwinist who "expounded the idea of struggle for survival into a doctrine of ruthless competition and class conflict."[60] Yet others have argued that this view of Spencer as someone whose philosophy had a paralyzing effect on the will to social reform is very misleading.[61]

Although Spencer opposed state charity and "state meddlings with the natural play of actions and reactions," he insisted that "it does not follow that the struggle for life and the survival of the fittest must be left to work out their effects without mitigation."[62] He maintained that aid for the "inferior" should be supplied by the "superior" on a voluntary basis and thereby kept within moderate limits to the "benefit of both—relief to the one, moral culture to the other. And aid willingly given (little to the least worthy and more to the most worthy) will usually be so given as not to further the increase of the unworthy."[63] As long as competition and "that natural relation between merit and benefit" were maintained, there would be an increasing integration of all members of society into a differentiated, mutually dependent, and efficient higher *social organism* in which violent competition (such as war) was replaced by the "peaceful competition" of the free market.

Spencer was committed to individualism and to unfettered capitalism; some writers exploited the apparent contradiction between his individualism and his organismic view of society. For example, socialist reformers such as Beatrice Webb and the reform-minded American sociologist Lester Frank Ward argued that Spencer's metaphor of "the social organism" (which Ward dismissed) would imply the subordination of the individual to the needs of the whole and the need for the state as the "organ of integration" to coordinate individuals into the social organism.[64]

Without entering into debates about the extent to which Spencer's philosophy impeded a will to reform, it is enough to emphasize here that for Spencer no less than Huxley, natural selection and the struggle for existence provided a plausible explanation for all the selfish behavior of which man is capable.[65] But all was not a gladiatorial show for Spencer; cooperation and altruism were widespread in nature though secondary to and dependent upon egoism.[66]

In Spencer's "synthetic philosophy," evolutionary progress proceeded from the "homogeneous" to the "heterogeneous," from the "incoherent" to the "coherent." As it did for the human social realm, progress in the natural realm meant increased specialization and integration into an organic whole. In the 1899 edition of his *Principles of Biology,* with the help of the well-known plant ecologist Arthur G. Tansley, Spencer added a new chapter, "The Integration of the Organic World," "to round off the general theory of Evolution in its application to living things."[67] The increasing integration among plants and animals, he argued, demonstrates "the law of Evolution . . . under its most transcendental form."[68] Symbiosis, which he identified with mutualism, offered what he considered to be crucial evidence for his organismic view of nature.[69]

Symbiosis for Spencer was a phenomenon which was to be expected and understood in terms of the physiological division of labor. The principle of division of labor was borrowed from the politicoeconomic theory of Adam Smith. As Camille Limoges has shown, it went through a series of transformations under the pen of biologists. The French zoologist Henri Milne-Edwards developed the concept "physiological division of labour," which Darwin subsequently used in his theory of species divergence, called by

Limoges a "division of ecological labour."[70] In *The Origin,* Darwin wrote that "the advantages of diversification of structure in the inhabitants of the same region is, in fact, the same as that of the physiological division of labour in the organs of the same body."[71]

Spencer took the analogy literally: the physiological division of labor actually represented the differentiation of a superorganism. He argued that while Darwin and his followers had been occupied with explaining the genesis of variations leading to new species, there had been "little if any recognition of an accompanying change, no less fundamental. In the general transformation which constitutes Evolution, differentiation and integration advance hand in hand; so that along with the production of unlike parts there progresses the union of these unlike parts into a whole."[72]

Mutual dependence and integration occurred early in the evolutionary progression of life proposed by Spencer. In the earliest primordial forms of life which possessed a "homogeneous nature," and in the absence of kinds, there was no "mutual dependence." But as fast as the original types differentiated into organisms with traits that were more markedly vegetal and others more markedly animal, Spencer argued, there came the beginnings of cooperation: the vegetable types by the aid of light formed organic matter from the inorganic world, and the predominantly animal types utilized the matter so formed. "Evidently with the rise of such a differentiation came an incipient mutual dependence."[73]

"Speculation aside," Spencer asserted that the distinction between Protozoa and Protophyta "foreshadowed that widest contrast which the higher organic world presents—the contrast between plants and animals."[74] The mutual dependence between these two great divisions was obvious: plants produce oxygen, animals inhale it and give back carbon dioxide, "they act reciprocally, as also in some measure by interchange of nitrogenous matters."[75] This "division of organic functions," he argued, was "the fundamental one which more than all others binds organisms at large together."[76] The phenomena of symbiosis provide him with "unexpected defense" for his view of life as an organic whole: "These present various cases in which the plant-function and the animal function are carried on in the same body,—cases in which the cooperation is not between separate vegetal organisms which accumulate nutritive matters and separate animal organisms which consume them, but is a cooperation between vegetal elements and animal elements forming parts of the same organism."[77]

Spencer referred to the fungi and algae forming lichens as a "communistic arrangement," which grew up after these two main types had become adapted to different conditions of life and had acquired appropriate specialties enabling each to benefit from what the other had acquired.[78] The presence of "vegetable cells" (algae) in amoebae, in numerous ciliated and flagellated *Infusoria,* in *Hydra viridis,* in various turbellarians, all illustrated, for Spencer, the same principle. Each of the partners profits. "Here, then," he declared, "we have exchange of services between associated plant-elements and animal-elements—a performance by them of different organic functions for the benefit of the aggregate which they unite to form."[79] These cases not only provided the lens through which to see how plants and animals were "mutually dependent parts of a whole"; they also prepared one to see how other kinds of relations between organisms which made them subservient one to another also constituted elements in the "general integration of the organic world."[80] Thus Spencer discussed van Beneden's commensalism as well as parasitism.

From a top-down perspective of a general whole created by the parts, Spencer could see mutual benefit everywhere. Even parasitic and predator—prey relations were not solely one-sided. One only had to look at different levels of organization to see benefit and to look "beyond immediate results" to see "certain remote results that are advantageous." They brought about changes which "though injurious to the individual are beneficial to the species, and which, when not beneficial to the species, are often beneficial to the aggregate of species."[81] Predators prevented "the inferior individuals—the least agile, swift, strong, or sagacious— . . . from leaving posterity and lowering the average quality of their kind." They saved individuals that were feeble by injury and old age "from suffering prolonged pains." All was for the common good, "the evils of death by disease and starvation being thus limited to the predatory animals, relatively small in their numbers."

At the same time, predators put a check on undue multiplication. Weasels not only benefit the plants eaten down by rabbits, and all those other animal which live on such plants, but also benefit rabbits themselves, since if they were to increase beyond their means of subsistence, a large part of them would, if not killed, die of hunger.[82] These examples illustrated the numerous bonds by which lives of organisms were tied together by mutual dependence so that one could "recognize something like a growing life of the entire aggregate of organisms in addition to the lives of individual organisms—an exchange of services among parts enhancing the life of the whole."[83]

As is well-known, other early leaders of plant ecology, Arthur Tansley in England and Frederick Clements in the United States, developed Spencer's superorganism concept. They maintained that competition among plants resulted in a highly developed division of labor in some plant communities, thus producing a more integrated and differentiated adult state.[84] This idea was the basis of the notion of succession in plant communities. As Clements put it, "The life-history of a formation is a complex but definite process, comparable in its chief features with the life-history of an individual plant."[85] Nonetheless, as ecology became focused on plant communities and animal communities, with little attention to 'underground ecology,' increasingly less attention was paid in ecology texts of the early twentieth century to "mutualistic symbioses" such as mycorrhiza, lichens, and algae living inside "lower animals."

Symbiosis as Parasitism

Those early plant ecologists who did discuss the phenomena of symbiosis did not account for it in terms of mutual aid brought about by a physiological division of labor among complementary types becoming integrated into an organic whole, as Spencer had suggested. They argued that mutualistic symbiosis, if it ever occurred, was extremely rare, and that interpretations of such relations in terms of cooperation for mutual benefit was mere sentimentalism. Examples of such cooperation could be found in some areas of the Animal Kingdom. But it was unreasonable and absurd to believe that "unconscious mutual support" existed among "lower" organisms such as algae, fungi, sponges, and ciliated protozoa, or among plants generally.

"In the plant community egoism reigns supreme."[86] This was the message of the Danish plant geographer Eugenius Warming, one of the European leaders of the new science of ecology. "The plant community," he declared, "is the lowest form; it is merely a congregation of units in which there is no cooperation for the common weal but rather a ceaseless struggle of all against all."[87] In the history of ecology, Warming is recognized for producing "the key synthesis that forced the scientific world to take note at last of the new field."[88] His classic work, *Plantesamfund,* was first published in 1895 and then in 1909 was revised and translated into English as *Oecology of Plants: An Introduction to the Study of Plant-Communities.* Although Warming's book was devoted to an understanding of plant communities, discussions of "symbiosis" occupied merely 9 of the 373 pages of the main body of the text. Even then, he used the term symbiosis to embrace the various kinds of bonds of different strengths—from the intimate to the loosest and most casual—that held individuals and species together to form communities.

In a section entitled "The Communal Life of Organisms," Warming discussed the relations of parasite with host, master with slave, mutualists, epiphytes, and saprophytes.[89] He interpreted all intimate associations in terms of parasite—host interactions and doubted that mutualism characterized by complete reciprocity of benefit between plants ever existed. In Warming's view, the lichen represented the exact opposite to mutualism. It was a case of slavery. He classified it as "helotism," after the system of serfage that had prevailed in Sparta: "The algae is in a condition of slavery in relation to the fungus, which is a kind of parasite differing from ordinary parasites in incorporating the host and in providing a portion of the food consumed in the host maintenance."[90]

In the few paragraphs devoted to mutualism, Warming mentioned mycorrhiza, the symbiosis of legumes and bacteria, blue-green algae living in special cavities in the underside of the leaves of azolla, and cyanophycae in the roots of cycads. But in his view such apparently stable and cooperative relations were exceptions to the normal behavior in plant communities. The principles for understanding the complex relations between species, he argued, had been laid down by Darwin.[91] They were never "at peace with one another." Every species "endeavoured" to extend its area of distribution by any means of migration it possessed.[92]

For Warming there was little complementarity in the sense of one plant producing deficiencies of others. They did not unite and combine their resources. Such cooperation was restricted to human and some animal societies.

The best plants could do for one another was to protect others from the elements, for example, when a larger shrub serves to shelter a smaller one from the wind. The only significant resemblance between animal communities and plant communities, he argued, was competition for food, which caused weaker individuals to be more or less suppressed.[93]

Similar views were maintained by the American botanist Roscoe Pound. In the late nineteenth century, Pound and his co-workers, including Charles Bessey and Frederick Clements, were engaged in an effort to introduce experimental methods and quantification into botany.[94] After collaborating with Clements on the phytogeography of Nebraska, Pound gave up science and went into law. Eventually he became dean of the Harvard Law School and a founder of the sociological school of jurisprudence. Pound was a Quaker; he opposed the welfare state but welcomed the cooperation of churches and fraternal organizations with the courts, saying: "This cooperation of organized religion and organized morality with the law is the more gratifying, because if individual, self-reliant, free enterprise has been an American characteristic *cooperation* has not."[95]

In his paper "Symbiosis and Mutualism" (1893), Pound attempted to dissociate the two terms which had often been confused: "While mutualism, in the case of plants, can only exist with symbiosis, in the larger portion of cases of symbiosis there is no mutualism."[96] All apparent examples of cooperation for mutual advantage were simply that—apparent. Mutual cooperation did not emerge among divergent microbes early in evolutionary history. Microbes were parasites. They appeared to be cooperative only when one looked superficially at the end result of some symbioses. In mutualistic arrangements, he argued, one of the associates always dominated, and if the other associate species benefited, it did so at a price to individual members of the species. It was therefore crucial, Pound argued, to understand the cost as well as the benefits accruing to the associates, and to distinguish between individuals and populations when discussing mutualisms between species. For example, wheat is cultivated by humans and enabled to grow in quantities and in localities which, under ordinary conditions, would be impossible. However, "it gains this partial exemption from the struggle for existence only at the expense of an immense number of individuals sacrificed, but it is, nevertheless, a great advantage which it gains. This may be called mutualism."[97] Pound made similar arguments for lichens. If the colony of algae benefited by protection, it was at a cost of many lives of individual algae that were attacked by the fungus.[98]

Pound was especially critical of Frank's claims and interpretations of symbiosis in terms of mutualism. He dismissed them as "decidedly fishy."[99] For example, Frank had suggested (as did de Bary) that some species of algae may have become so adapted to life in lichens and so accustomed to it that they partially or wholly lost the power of independent growth. But Pound rejected any suggestions that such cases could result in perfect reciprocity leading to the formation of complete, inseparable interdependence. He emphasized that no examples of dependent algal species were known, and he did not expect them to be found.[100] Pound also denounced Frank's claim that mycorrhizal fungi benefited trees and gained nothing in return. As he saw it, this was merely one of "Frank's statements calculated to try our patience and credulity."[101] Instead, he supported the claims of more "sober" writers who argued that the fungi did no benefit to the trees and orchids they infect. Indeed, he asserted that they were "probably injurious by taking nourishment properly belonging to the tree."[102] Finally, Pound repudiated Frank's suggestion that some secretion gave the roots of the legumes the power to attract the bacteria. Bacteria were parasites. "They are there for their own purposes, and are incidentally beneficial to the plant."[103]

The metaphor "mutualism," borrowed from political movements, opposed another metaphor, "inheritance," borrowed from jurisprudence. In the sense of hereditary succession to property, rights, and the like, the concept of inheritance went through various transformations in the hands of biologists. As developed by the master metaphorician, Darwin himself, it referred to those characters that could be attributed to inheritance from a common progenitor.[104] But as it pertained to infectious microorganisms, the message was always the same: microbes were *thieves,* there to steal from the "host" its *rightful inheritance.* As Pound put it,

> It is not necessary, as Frank seems to think, in order to establish mutualism to show that the organisms do no injury to each other. Mutualism of the kind we meet with in the vegetable kingdom involves sacrifice on the part of the host. The parasite is not there gratuitously. It is there to steal from its host the living it is hereditarily and constitutionally indisposed to make for itself. If the host gains any advantage from the relation, it can only do so by sacrificing—by giving the parasite the benefit of its labor that it may subsist.[105]

Pound's argument implied that those who observed such "mutualistic symbiosis" at the level of species had simply failed to recognize the underlying evolutionary mechanism of individual conflict and domination, as well as the parasitic nature of microbes. This line of argument continued throughout the twentieth century, as many biologists were averse to the word symbiosis if it meant that different species "harmonize their functions for the greater good of the community." However, not all biologists around the turn of the century came to believe that *all* cases originated as infectious disease. Moreover, many continued to define symbiosis in terms of relations between species, viewing them as entities as opposed to populations of individuals.

The Synthesis of New Individuals

To the few botanists who investigated mutualistic symbiosis among 'lower plants and animals' and believed that cooperation between such phylogenetic groups was of wide occurrence, it was clear that symbiosis could lead to highly specialized morphological and physiological changes and result in new and "higher" organic wholes. Indeed, some believed it possible that this process had led to the evolution of *all* plants and animals. Lichens, mycorrhiza, bacteria in the root nodules of legumes, and algae in animals provided the clues. This argument was most highly developed by the American lichenologist Albert Schneider at the University of Illinois.

In 1897 Schneider attempted to systematize the phenomena of symbiosis both in his *Text-book of Lichenology*[106] and in a general overview simply entitled "The Phenomena of Symbiosis"[107] published in the first volume of the journal *Minnesota Botanical Studies.* The textbook offered a detailed history of lichenology, reviewing the work of European botanists concerning the origin and classification of lichen. Schneider favored the views of Johannes Reinke and his followers that lichens represented a new class of their own and that they had a polyphyletic origin resulting from different algae living in combination with different fungi. His essay on the phenomena of symbiosis represented a synthesis of such views.

Schneider began with a new definition of symbiosis that did not conflate the phenomena of symbiosis with the complex interdependence in communities or societies which de Bary had proposed. He acknowledged that etymologically the word symbiosis signified a living together and was therefore perfectly fit for use in the broader sense. But in Schneider's view there was a great difference between the mere association of living things and what constituted "true symbiosis." Although both societies and symbiosis were evidence of biological interdependence, the general difference was that in the former the interdependence was remote, whereas in the latter it was very close, involving a physiological relationship. This physiological relationship, in turn, entailed the organisms being brought into intimate contact. Thus he redefined symbiosis as "*a contiguous association of two or more morphologically distinct organisms, not of the same kind, resulting in a loss or acquisition of assimilated food-substances.*"[108]

While this definition excluded phenomena associated with "sociality," it included many other kinds of phenomena hitherto not classified as symbiosis. As Schneider remarked, "A mere mention of all the experiments and discoveries in connection with symbiosis would fill volumes": the recent development of germ theory and discoveries in the treatment of disease, surgery, agriculture, and dairy industries were all directly concerned with some form of symbiosis.[109] The relationship of male and female reproductive cells represented a most specialized form of mutualistic symbiosis where one organism was created out of two. The relationship between the immature embryo and mother represented another form of symbiosis: parasitism.[110] Grafting was also of a symbiotic nature. The functional nature of graft and stock formed "a most perfect symbiosis": mutualism.[111] Tumors (sarcoma, carcinoma, etc.) and cysts of various kinds were also included under his definition of symbiosis. Though the origin of these growths was little understood, he argued that they were nevertheless foreign to the body in which they live as true parasites.[112] Contemporary investigations and discoveries in regard to immunity, toxins, and antitoxins were also aspects of the study of symbiosis.[113]

But Schneider was most concerned with the phenomena of symbiosis which already had been recognized as such. He classified symbiotic associations in terms of antagonism and various kinds of mutualism, and he theorized on the kinds of morphological and physiological changes that would result from them. In "antagonistic symbiosis," or parasitism, the morphological and physiological specializations and adaptations were limited. One of the symbionts may be highly benefited while the other was always injuriously affected. It was therefore a destructive association: the changes tended toward "dissolution rather than evolution"—from "the higher to the lower."[114] To illustrate how antagonistic symbiosis led to dissolution, he offered an account of the origin of parasitic fungi. It was based on the generally held belief that they were derived from algae that had lost their chlorophyll function.[115] Owing to a lack of space due to overproductiveness, certain algae had frequently come in contact with more complex plants and animals from which they absorbed various organic food substances. This contact reduced the need for chlorophyll assimilation in the algae, resulting in a corresponding change in structure. As the morphological and physiological changes in the algae moved in the direction of parasitism, the algae finally lost their chlorophyll function and depended solely on an organic food supply. The result was the origin of fungi. In Schneider's view, it was possible that the majority of symbioses were originally more or less antagonistic. He was certain that yeast was originally green algae, and it was still a matter of dispute whether most bacteria were derived from algae.[116]

Mutualistic symbiosis was far more important for "progressive" evolutionary change. As Schneider conceived of it, each symbiont possessed or developed a specific character which was useful for the other symbionts. The morphological changes that accompanied the functional relationships could be very marked or scarcely perceptible, and the adaptation was neither quantitatively nor qualitatively equal for all the symbionts. "Theoretically," he remarked, "there is no limit to the degree of specialization and perfection that this form of symbiosis may attain. In fact, mutualistic symbiosis implies that there is a higher specialization and greater fitness to enter into the struggle for existence."[117] Lichens illustrated this beautifully. These plants, Schneider argued, were of wider distribution and possessed greater vitality and physiological activity than either of the symbionts. They occurred in the tropics as well as the extreme north; in the lowest valleys as well as on the highest mountain peaks. Indeed, he asserted that their vitality was greater than that of any other plants.[118] The root tubercles of legumes were a further example of an evolutionary morphological change resulting from mutualistic symbiosis. The tubercles, he argued, were "neoformations" induced by the bacteria which grow and multiply in the parenchymal cells. The bacteria took their food supply directly from the cell contents of the plant. In exchange, the plant received nitrogen compounds formed by the bacteria in the process of binding the free nitrogen of the air.[119]

For "one-sided mutualism," Schneider coined the term "nutricism": "a form of symbiosis in which one symbiont nourishes the second symbiont without receiving any benefit in return."[120] Mycorrhiza was exemplary. Schneider accepted Frank's interpretation that the function of the fungus was to supply the tree with food substances and moisture taken from the soil. He asserted that it had been proved that the tree is greatly benefited, while no evidence could be found to indicate that the fungus is benefited. The case of algae living in tubercular outgrowths from the roots of cycads, which he himself had investigated, provided a further example of nutricism.[121] In this case the algae benefited from the host for food supply, but he could find no evidence to show that the host was benefited or harmed in any way.

Cases of mutualistic symbiosis in which both associates could lead independent existences were distinguished from those in which one or all of the symbionts were permanently dependent. The latter, Schneider designated as "individualism": the associations formed an individual, a morphological unit, and the phenomena were frequently not recognized as symbiosis.[122] "Higher lichens" represented the best known and perhaps the most typical form of complete individualism. Schneider admitted that most botanists were agreed that while the fungal symbiont had entirely lost the power of independent existence, the alga could exist independently. Nevertheless, like Frank and de Bary, he believed that in some cases at least, the algae may likewise have lost the power to continue independent existence. Based on evidence of these kinds of transitional stages, he prophesied that

> future studies may demonstrate that the cell, and hence the individual, is neither more nor less than complete individualism. The plasmic bodies, such as chlorophyll granules, leucoplastids, chromoplastids, chromosomes, centrosomes, nucleoli, etc., are perhaps simply the symbionts comparable to those in the less highly specialized symbiosis. Reinke expresses the opinion that it is not wholly unreasonable to suppose that some skilled scientist of the future may succeed in cultivating chlorophyll-bodies in artificial media.[123]

The suggestion that such cellular bodies might be symbionts was not the unique insight of Schneider standing alone in his laboratory in Champaign, Illinois. In the late nineteenth century such ideas peppered discussions of the nature of cell organization.

3

Socially Constructing the Individual

> It is but a short step from this conclusion to the view that the centrosome, too, is such an independent organism and that the cell is a symbiotic association of at least three dissimilar beings! Such a conception would, however, I believe, be in the highest degree misleading. . . .
>
> E. B. Wilson, 1896[1]

It was often proclaimed in the late nineteenth century that the key to all biological problems must, in the last analysis, be traced to the cell.[2] The discovery that all cells arise by division of preexisting cells, neatly embodied in Rudolf Virchow's maxim "omnis cellula e cellula"; the extension and verification of this maxim in the demonstration that the vertebrate egg is a single cell; and the demonstrations of the 1870s and 1880s that the internal processes of cell division were fundamentally the same in both plants and animals—all magnified the importance of the cell as a universal unit of structure. Cell theory, inaugurated a new era in the history of physiology and pathology based on the belief that all the various functions of the body, in health and disease, were but outward expression of cell activities. It was through cell theory again that the way was opened for understanding the nature of embryonic development and the law of genetic continuity that lay at the basis of inheritance. It was frequently claimed that no other biological generalization, save only the theory of organic evolution, had brought so many apparently diverse phenomena under one common point of view or had accomplished more for the unification of knowledge.

Symbiosis offered another point of unification. In the late nineteenth century, evidence was accumulating for the existence of various cellular bodies which possessed properties of growth, assimilation, and division. When associated with the cases of lichens and of algae living in "lower animals," this evidence led several biologists—zoologists as well as botanists—to consider the possibility that the cell, and hence the individual, was a symbiosis of various self-reproducing bodies of different phylogenetic origin. At this juncture, with cell theory, discussions of symbiosis shifted and came to rest chiefly on evidence for the physical continuity of cell constituents and their status as relatively autonomous organismic entities. That the complex organism was made up of smaller independent and interdependent elementary organisms—cells—was a basic tenet of the cell theory of development. It was not a great leap to suggest that cells themselves were similarly made up of still smaller mutually interdependent organisms. Indeed, at first glance the symbiotic origin and nature of plants and animals might appear to have been in step with contemporary concepts of the organism.

Cell Theory Is Social Theory

Historians of biology have devoted considerable attention to the origin of cell theory, the study of microorganisms, the establishment of marine biological laboratories, and the rise of experimental embryology. The focus of this work initially lay in Germany. Some have argued that cell theory first emerged from the German Romantic tradition, with its strong commitment to synthesis to bring together a large number of phenomena within a unified framework.[3] Others have emphasized technical improvements in microscope construction with the rise of the precision optical industry in Germany; the availability of aniline dyes and the discovery of their use in cytology; the need for adequate medical care, as perceived by the state, with an underlying base of medical research.[4] Not only did social interests stimulate research in cell and developmental biology, but social relations were embedded in the very concepts of cell and developmental biology of the nineteenth century.

Cell theory itself was a social theory as expressed first in the theory of the "cell-state" or "cell-republic" elaborated by Virchow, Haeckel, Oscar Hertwig, and many others, and second in symbiotic theories of intracellular organization. Both of these theories rose to the center of considerable debate and framed the way in which biologists approached problems of organismic organization. One can find a common basis for cell theory and social theory in the late nineteenth century in the same way it has been shown for theories of evolution.

Biologists generally regarded a cell as a living, individual organism. All distinctive vital processes—metabolism, growth and reproduction, sexual phenomena, and heredity—reduce themselves ultimately to activities taking place in, and carried on by, the individual cells. The cells composing the complex bodies of animals and plants were understood as mutually interdependent, and, with the exception of the mature germ cells, could not maintain their existence apart from their fellows. That is to say, the only natural environment suitable for their continued existence was the complex body or "cell commonwealth" of which they formed an integral part. According to this view, the only difference between Metazoa and Protista was that in the simplist forms of life the whole body of the living individual could, for whatever reason, reach no higher degree of complexity than the single cell. All higher organization was supposed to evolve through the principle of "physiological division of labor," reaching its fullest expression in the mutuality of the constituents.

The predominant way of understanding the role of cells in development thus began with the concept of cells as individuals and the principle of division of labor. The development of a cellular organism was conceived of in social terms. A simple colony of like cells evolves into a commonwealth of differentiated and mutually dependent cells. In proportion as division of labor is carried out, interdependence is increased, and the individuals become more and more intimately associated. A multitude of independent individuals, adopting mutual service as the best economy, find themselves in the end so firmly bound together in interdependence that they constitute a complex individual. The struggle for existence was supposed to extend to cells and even to intracellular entities. Herbert Spencer in 1893 stated the usual conception of this division of labor:

> An *exchange of services,*—an arrangement under which, while one part devotes itself to one kind of action, and yields benefits to all the rest, all the rest, jointly and severally performing their special actions, yield benefits to it in exchange. Otherwise described, it is a system of *mutual* dependence.[5]

Leading cell biologists in Germany, including Virchow and Oscar Hertwig, stressed that, as biology dealt with the organization of life, it was more akin to the social sciences than the physicochemical sciences.[6] Paul Weindling has argued that the theory of the cell state was used by leading biologists in Germany to support a *variety* of social theories about how to run the state—from the most progressive to the most reactionary. But organicist theory was not used merely to prescribe larger societal relations. German biologists also used it as rhetoric in their attempts to unite the sciences of anatomy, physiology, zoology, and botany around cell theory.[7]

These kinds of statements echoed across the Atlantic, where one could hear C. O. Whitman, first director of the Marine Biological Laboratory at Woods Hole,[8] stating in 1891 that naturalists were long accustomed to the idea that the living body represents "a commonwealth of cells." The metaphor, he argued, was based not on "superficial or fanciful resemblances" but on "analogies that lie at the very foundation of organic and social existence."

> On the same grounds that the sociologist affirms that a society is an organism, the biologist declares that an organism is a society.
>
> A society is an organized whole, the unity of which consists in, and is measured by, the mutual dependence of its members. The living body is an organization of individual cells with the same bond of unity. The principle of organization in both cases is the division of labor or function.[9]

Whitman declared that the principle of division of labor underlay all organic as well as social progress.[10] It was as important for the organization of biology as it was in the biology of organization. The days when naturalists could presume to take all nature for a subject of study and meditation were long gone. The cosmogonist of olden times had engaged single-handedly with all the mysteries of the universe. Whitman asserted that one could honor them for their heroic efforts, but they had been misdirected and ineffective. At the expense of centuries of baffled efforts the lesson began to be learned that division of the problem facilitates progress. Though a thing of slow growth, "self-originating, self-perpetuating and self-regulating" division of labor had "taken possession of the biological sciences, and presides over their onward march, just as it determines and directs social and industrial progress."[11]

But Whitman recognized that specialization in itself could never lead to progress. Following Spencer, he argued that progress was found in "complete separateness and complete union," when the "highest individuation is joined with the greatest mutual dependence."[12] This meant that one needed to bring together "scattered forces into something like a union." Thus he advocated the foundation of a "national marine biological station" that would bring together the largest number of the leading naturalists of the country and place them "in intimate helpful relations" to one another. "The larger the number of specialists working together, the more completely is the organized whole represented, and the greater and the more numerous the mutual advantage."[13]

The problem of organizing ever-increasing specialization in the life sciences into a coherent whole reverberated in concepts of the cell and ontogenetic development. In the first edition of *The Cell in Development and Inheritance* (1896), the American cell biologist E. B. Wilson declared that there was "no biological question of greater moment than the means by which the individual cell-activities are coordinated and the organic unity of the body maintained."[14] He asserted that the theory of the cell state had taken a very strong hold on the minds of biologists, and it was widely accepted.[15] In the third edition of *The Cell* (1925), he continued to celebrate the conception of the cell state, claiming that, as "elaborated by Milne-Edwards, Virchow and Haeckel, this conclusion offered a simple and natural point of attack for the problems of cytology, embryology, and physiology, and revolutionized the problems of organic individuality."[16]

The Cell as a Collective

The idea that an individual was made up of smaller individuals, each possessing the properties of growth, assimilation, and reproduction, was applied to every grade of biological organization. If plants and animals were made up of smaller individuals—cells—they in turn might be made up of still smaller "elementary organisms." During the second half of the nineteenth century, many leading theorists postulated that underlying the structure of the cell there existed microscopically invisible living units standing somewhere between the cell and the ultimate molecules of living matter. These living units, or hypothetical "elementary organisms," were the starting point of every leading theory of heredity and development. The "physiological units" of Herbert Spencer, the "gemmules" of Charles Darwin, the "biophors" of August Weismann, and the "pangenes" of Hugo de Vries were widely discussed during the late nineteenth century. Though these hypothetical entities were not synonymous, all of them were endowed with powers of assimilation, growth, and division, and all were thought to somehow build up cells and complex organisms. They were postulated as logically necessary but had little other basis.

Nevertheless, cell biologists generally came to accept that the ultimate basis of living matter was a mixture of many chemical substances that were self-perpetuating without loss of their specific character. The open question was whether these substances were localized in discrete morphological bodies (aggregated to form the cell somewhat as cells were aggregated to form a multicellular organism) and whether such permanent bodies, if they existed, were within the reach of the microscope. This was generally held to be true for at least one part of the cell—the nucleus, which consisted of self-reproducing units of a lower order than itself, the chromosomes. With the aid of the light microscope coupled with the use of chemical substances that selectively stained cellular structures, it became apparent by the 1890s that protoplasm was not merely a bag of complex chemical compounds. One could see an abundance of various rods, threads, membranes, vacuoles, pigment bodies, and other granules, though the most obvious of these parts was the cell nucleus.

It was generally recognized that with the possible exception of some of the lowest forms of life, such as bacteria, all cells contained a *nucleus*. Thus the cell came to be defined morphologically as a mass of protoplasm containing a nucleus. This cell structure was held to be the common basis of plants and animals. Within the nucleus, the darkly staining chromosomes began to assume the chief role in providing the physical link between the cell and evolution.[17] Their physical continuity through the cell cycle, the constancy of their numbers, the accuracy of their movements, and the longitudinal splitting of the chromatin threads, along with the fusion of male and female pronuclei in fertilization, all combined to give them an exceptional position in cytological discourse. Like the cell, the nucleus exhibited physical continuity; it was never formed de novo, but always arose by division of a preexisting nucleus.

There was also a definite chemical contrast between the nucleus and the protoplasmic substance surrounding it, the *cytoplasm*. The former was characterized by the abundance of a substance rich in phosophorus known as *nuclein,* while the latter was believed to contain no true nuclein and was rich in protein and related substances. The differentiation of the protoplasmic substances into nucleus and cytoplasm was generally held to be a fundamental character of the cell, both in a morphological and in a physiological sense. The nucleus was often regarded as the "controlling center" of cell activity, and hence a primary factor in growth, development, and the transmission of specific qualities from cell to cell and so from one generation to another. The cytoplasm was often referred to as the "cell body." The structural and chemical differences between nucleus and cytoplasm were of so marked and constant a character that not only were they regarded as the most important of all cell differences, but to some the nucleus appeared to be a perfectly distinct body suspended in the cytoplasm.

The history of this cytological work usually ends here, and historians generally argue that by the mid-1880s the nucleus came to be regarded by leading biologists as the fundamental cellular element, at once ensuring heredity, allowing variation, and directing ontogeny. The Mendelian chromosome theory of inheritance developed during the second decade of the twentieth century would be the ultimate offspring of this work. However, looking for the precursor or physical basis upon which the Mendelian theory rested results in a deceptive history. If we reel the nineteenth century film forward a few more years and look around, we see that not all biologists believed that the nucleus was the sole bearer of hereditary qualities, or even the most important center of formative substances.

Between 1886 and 1896 the histologist Richard Altmann (1852–1901) at Leipzig described (by means of an elaborate technique) certain cytoplasmic granules which he believed to be in the cells of all animals. These granules had often been reported in many kinds of cells and gave the cytoplasm its grainy appearance. Some had believed them to be inert particles produced by the living protoplasm and pulled along mechanically by protoplasmic movements. But Altmann bestowed the properties of life on these granules and named them "bioblasts": they were "elementary organisms" which lived in a homogeneous fluid, or ground substance of the cytoplasm.

Elaborating this concept in his treatise *The Elementary Organisms and Their Relationship to the Cell* (1890, 1894), Altmann attributed to bioblasts the most important and most general functions of life, in particular, the secretion of various cell substances, fat, glycogene, pigments, and the formation of fibers in nerve cells.[18] Bioblasts were of diverse nature; each kind had properties of its own. The cell then would not be the elementary organism, the indivisible unit of life; it would yield this role to bioblasts and would be itself only a colony of bioblasts. Thus the theory of genetic continuity expressed by Rudolph Virchow in the aphorism "omnis cellula e cellula" appeared in the writings of Altmann as "omne granulum e granulo!"[19]

Altmann offered a brief account of cell origins, suggesting that cells first emerged when originally separate bioblasts came together into a colony. It was well-known that certain bacteria come together and secrete an enclosing envelope or capsule of gelatinous or gummy material. This mass of bacterial cells, called zoögloea, gave Altmann some idea of what this stage of evolution would have been like. In zoögloea it was the particles which were living, and the intermediate substance was only an inert product. The resemblance was close, but the protoplasm of cells had evolved beyond the zoögloea stage. A membrane was constructed which isolated and individualized the bioblast colony. Thus the first nonnucleated cell was formed. The last stage of this process resulted in the formation of the nucleus.

Altmann's theory of bioblasts was criticized from various perspectives. The renowned French biologist Yves Delage[20] applauded the theory for being based on real entities which everyone had seen and for being therefore much less hypothetical than all previous theories of elementary organisms. Delage did not reject Altmann's origin story based on nomadic individuals settling down and forming a wall around themselves. On the other hand, when measured against the major problems of the day—heredity, ontogeny, variation, and adaptation—he argued, Altmann's theory did not go far enough. Moreover, it seemed to have several serious logical problems with respect to the mechanism by which bioblasts could be transmitted to successive generations.

Other cytologists were much more severe in their criticisms and rejected the very facts upon which the theory was based. They denied the real existence of the granules, dismissing them as merely artifacts produced by the coagulating effects of the staining reagents Altmann used.[21] But Altmann's claims for the existence of such granules were vindicated. In 1897 the German histologist A. Benda developed another complicated technique which permitted more detailed examination of these granules. He renamed them "mitochondria" from the Greek *mitos* (thread) and *chrondos* (granule). With the further refinement of cytological techniques, studies of mitochondria emerged as a prominent field, especially in Germany and France during the first decades of the twentieth century.

In the meantime, another intracellular entity, which Theodor Boveri (1888) named the "centrosome" (and later the "centriole"), attracted attention. The centrosome was thought to be present in most, if not all cells and to be especially concerned with the processes of cell division and cell reproduction. In 1887, Boveri and Edouard van Beneden (today frequently confused with his father Pierre-Joseph) independently suggested that the centrosome was an independent permanent cell organ which multiplied by division to form the centers of the daughter cells. The important role of the centrosome in presiding over cell division led van Beneden to believe that it was "equal in rank to the nucleus itself" and Boveri and his followers to regard it as the "dynamic centre of the cell."[22] Nonetheless, as late as the 1930s, some biologists thought that centrosomes, too, were merely artifacts induced during the preparation of cells for microscopy.[23]

By the mid-1880s many cytologists came to believe that *plastids* were capable of growth and division and were transmitted like the nucleus from cell to cell and from organism to organism. The most important of these were the *chromatophores,* the colored plastids including the "chlorophyll bodies" which Andreas Schimper (1856–1901) had named *chloroplasts* (*chloroplastiden*). Schimper's studies were generally regarded as the first convincing demonstration that these cell "organs" arose only by division of preexisting bodies of the same kind. In his lengthy and classic paper of 1885 on the structure, physiology, and continuity of plastids, Schimper compared them to independent organisms.[24]

During the 1880s and 1890s, the major distinction between nucleus and cytoplasm and the evidence for self-reproducing centrioles and plastids was taken by several botanists and zoologists to imply that all cells (bacteria notwithstanding) might be composites of several phylogenetically distinct elementary organisms, each carrying out specific functions in a cell cooperative.

Schimper's studies of chloroplasts provided the exemplar. A former student of Anton de Bary, Schimper was well aware of the research on the dual nature of lichens and other examples of microbial symbiosis.[25]

Schimper actually had this idea in mind before he demonstrated that chloroplasts were self-reproducing. In 1883—in the same paper in which he coined the term chloroplast—he suggested that symbiosis was one of the theoretical stakes:

> Should it be definitively proven that the plastids are not formed anew in the egg cells, then their relationship to the organism that contains them would more or less remind us of a symbiosis. It is possible that the green plants indeed owe their origin to the union of a colorless organism with one evenly stained with chlorophyll.[26]

Schimper, who moved from Strassburg to teach botany at the University of Bonn, had detailed training in experimental physiology, particularly in photosynthesis and the metabolism of plants. However, he never developed the idea of plastid symbiosis. Instead, he later took a new ecological direction in his research and applied his laboratory background to the study of the physiological adaptations made by individual plants and whole societies to such external conditions as heat, precipitation, and soil.[27]

The analogy between symbiotic algae living in the cells of many translucent animals and the nature of chloroplasts in plants was obvious to many botanists. In some cases, the chlorophyll bodies contained in the colorless cells of some animals had been shown to have lost the ability to exist independently. One of the best known cases concerned the green cells of the aquatic flatworm (turbellarian) *Convoluta roscoffensis.*[28] Studies of this species by one of Simon Schwendener's students, Gottlieb Haberlandt, led him, in 1891, to support the suggestion that chloroplasts originated as symbionts. He believed that the green cells of *Convoluta* had become *organs* of the colorless organism; they were so greatly modified as to be barely recognizable as derived from an independent source. He suggested that they might undergo further modification to lose their own colorless protoplasm and nucleus, thus evolving into simple chlorophyll corpuscles like those of higher green plants.[29]

It was not long before this reasoning was applied to the nucleus, cytoplasm, and centrosome. In 1893, in a lecture given before the Biology Club of the University of Chicago, the Japanese zoologist Shôsaburô Watasé (1862–1929) offered the suggestion that these cell structures were symbionts. Watasé had visited the United States as a student in 1886 and completed his Ph.D. in 1890 under William Keith Brooks at Johns Hopkins University.[30] However, Watasé did not begin cytological studies until his association with C. O. Whitman. Between 1890 and 1899, he was a member of Whitman's staff, first at Clark University and then at the University of Chicago and at the Marine Biological Laboratory at Woods Hole, where he studied the cleavage of the ovum of squid.

Watasé became well assimilated into American social and intellectual life.

As some of his colleagues later recalled, "His friends had ceased to regard him as foreign, and thought of him as a permanent acquisition."[31] Yet he did not become a "permanent acquisition" to his host nation. In 1899 he decided to return to Japan and accepted the chair of zoology at the University of Tokyo. Nor did he continue his work on symbiosis or cytology. Instead, he became involved in organizations to preserve natural and historical monuments of Japan, established a society for the study of the ecology of animal life in eastern Asia, and wrote a popular book in Japanese on animal phosphorescence.

While in the United States, Watasé's work on the cleavage of the egg of the squid led him into studies of the centrosome[32] and the nature of cell organization. In his lecture on this subject, he presented the relation between the cytoplasm and nucleus of the cell as a symbiosis of phylogenetically distinct organisms. As late as 1930, the renowned American cell and developmental biologist F. R. Lillie could still claim that "this idea has in it much for the fruitful consideration of modern cytologists."[33] Watasé argued that one could study cell organs from three points of view: how the parts were adapted by form and structure to perform their physiological work; where and how they originate in the cell; and the probable steps in the ancestral history by which these structures came into existence. This last phylogenetic perspective belonged to what he called *cytogeny.* By extending the morphological study of an organ through organisms of different grades of complexity, one could infer the probable steps through which it may have passed in the course of its phylogenetic history.

Watasé listed Reinke, Schimper, and Haberlandt in support of the view that chloroplasts were, in a phylogenetic sense, to be regarded as algae. He also mentioned the classic papers which viewed "animal chlorophyll" as veritable algae.[34] All these cases, he argued, underscored the fact that certain parts of an organism, which had once been considered to be integral elements of the organism derived from the differentiation of the germ, were in reality due to an association of two or more different organisms. They all showed that what one called *organs,* from a physiological point of view, were in reality *organisms* by themselves.[35] But if this work left any room for doubt, he asserted, the phylogenetic history of lichens indisputably showed how an independent organism, composed of organs, could be created by the union of two dissimilar organisms by the establishment of an intimate physiological relationship between them.[36]

Watasé applied this argument to the cell cytoplasm and nucleus. Though he never referred to Altmann's bioblasts or Benda's mitochondria, he argued that the behavior of the cytoplasmic thread or network suggested that it was "formed of a group of small, living particles, each with the power to assimilate, to grow and multiply by division." The chromosome was itself "a colony of minute organisms of another kind," each endowed with similar attributes of vitality. That the cell as a whole assimilates, grows, and divides "is ultimately due to the fact that the minute particles which compose the cytoplasm and chromosome are endowed with these functions."[37] In his view, it had not been definitively established that the centrosomes were self-perpetuating or self-reproducing structures. But if this were shown to be the case, then the arguments he advanced would apply to them as well. Thus he confined his remarks to the phylogenetic origin of the nucleus and cytoplasm.[38]

The "differentiation hypothesis" for the origin of cell organs was upheld by several leading biologists,[39] but to Watasé, it was simply inadequate. There was no evidence proving that the nucleus was formed by a process of differentiation from the cytoplasm, or that the cytoplasm was formed by the differentiation of nuclear substance. Watasé maintained that during cell division and during fertilization, the identity of the nucleus and cytoplasm was never lost. He insisted that any theory of the organization of the nucleated cell had to recognize the profound physiological interdependence between nucleus and cytoplasm, their reciprocal interchange of metabolic products, and their morphological independence. "The doctrine of symbiosis, first propounded by De Bary, just fulfils [sic] these requirements, inasmuch as it means now, in a more restricted sense, *the normal fellowship or the consortial union of two or more organisms of dissimilar origin, each of which acts as the physiological complement to the other in the struggle for existence.*"[40]

But Watasé sought harder experimental proof for the symbiotic nature of the nucleated cell—of the kind botanists had provided for the dual nature of lichens. He searched for a demonstration analogous to the success of Bonnier and Stahl in reconstructing lichens by bringing algal and fungal elements together synthetically. He found it in the experiments of Max Verworn,[41] who had managed to remove the nucleus from one cell of a certain kind of radiolarian and transplant a new nucleus into it. Watasé asserted that the synthetically produced radiolarian not only survived but could not be distinguished from the organisms upon which no operation had been performed. If the synthesis of lichens was conclusive evidence for their symbiotic nature, then Verworn's experiments should be convincing evidence of the symbiotic origin of the nucleated cell.[42]

Watasé was not prepared to speculate on what the earliest living units which formed cells actually were. But in assuming their existence, he was already on safe ground since leading biologists agreed that such physiological units existed in the cell. Thus he rested the main postulate that the nucleated cell was "a complex of at least two kinds of organisms, different in their anatomical character, in their function, and in their origin."[43] Thus he claimed that symbiosis as the basis of cellular organization agreed in its broadest features with the idea expressed by Darwin in his celebrated hypothesis of pangenesis. It gave a more concrete meaning to Darwin's well-known passage: "Each living being is a microcosm—a little universe, formed of a host of self-propagating organisms inconceivably minute and numerous as the stars in the heavens."[44]

Watasé, of course, was not simply giving this statement more concrete meaning. By associating symbiosis with it he was also dramatically changing its meaning. That germ cells contain innumerable ultramicroscopic cell-forming bodies had been a prototype for many diverse theories of heredity and cellular organization. Some of them were constructed in virtual opposition to the symbiotic origin and complementary physiological adaptations of cellular constituents.

Nucleocentricism

Symbiotic concepts of the cell and organism remained on the fringe of biological thought throughout the 1890s. Leading biologists continued to deny that cells were aggregates of discrete morphological bodies, defending the integrity of the cell itself as the "elementary organism," the indivisible unit of life, which through its metabolic action performed all the characteristic operations of life. For some, the idea that the integrated and organized nature of the cell could have arisen by microorganisms living together was simply incomprehensible—indeed, absurd.[45] But whether one believed that all cell structures arose through the action of invisible germs, or that they arose through the action of the metabolic activities of the cell as a whole, the symbiotic interpretation of the cell confronted an overwhelming belief in the nucleus as "the ultimate court of appeal" in cellular activities.

Cytologists of the late nineteenth century generally recognized that it was possible that many cytoplasmic structures were self-reproducing from one cell generation to the next. Nonetheless, many—including Wilhelm Roux, Hugo de Vries, August Weismann, and Oscar Hertwig—still insisted that they were products of nuclear activity. In his book *Intracelluläre Pangenesis* (1889), de Vries postulated that the gemmules of Darwin, which he called *pangenes,* were contained in the nucleus and migrated into the cytoplasm step by step during ontogeny, thus determining cell structure and the successive stages of development.[46] This proposal was developed into an elaborate theory by Weismann in his celebrated book *The Germ Plasm: A Theory of Heredity* (1893).

The arguments Weismann put forward for the nucleus as the sole bearer of hereditary qualities were echoed by leading American geneticists throughout the twentieth century. The first argument might be called the issue of the common denominator. In "higher organisms," the sperm cell is many hundred times smaller than the egg cell. Yet, he reasoned, "we know that the father's capacity for transmission is as great as the mother's."[47] The nucleus provided a place for equal transmission of hereditary substance from both parents. Second, studies of fertilization seemed to indicate that the essential part of this process consisted of the union of the nuclei of the egg and sperm cells.[48] Third, observations on cell division showed that the nuclear complex possessed a wonderfully exact apparatus for distributing the chromosomes in a fixed and regular manner.[49] Finally, he appealed to the "economy of Nature": "This substance can hardly be stored up in two different places, seeing that a very complicated apparatus is required for its distribution: a double apparatus would certainly not have been formed by nature if a single one suffices for the purpose."[50]

Thus Weismann proposed that vital units pass into the body of the cell through the nuclear membrane and there form its parts and structures.[51] He did not specify exactly when the first vital units constructed the cell, nor was he concerned with the actual process of morphogenesis or assimilation. He confessed that he knew no more about these isues than he did about the behavior of biophors in the cell body in response to those which migrated into it from the nucleus during the course of ontogeny. When arguing how the orderly nature of development might be controlled by a hierarchy of nuclear entities, he suggested that the invaders or "immigrants" from the nucleus might struggle with those already present in the cell body; the weaker ones would be suppressed and eaten by the stronger ones.[52]

Although Weismann's theory of development was severely criticized,[53] the belief that the nucleus alone contained the hereditary material and controlled all cell processes continued to gain support. As E. B. Wilson put it in the first edition of *The Cell,* "Both [nucleus and cytoplasm] are necessary to *development;* the nucleus alone suffices for the *inheritance* of specific possibilities of development."[54] Wilson recognized the striking similarities between the chlorophyll bodies studied in animals and the chloroplasts of plants. However, he doubted the interpretation that the chlorophyll bodies in many Protozoa and some Metazoa (hydra, sponge, and some planarians) were distinct organisms living symbiotically in the cell. "This view," he declared, "is probably correct in some cases, e.g. in the Radiolaria; but it may well be doubted whether it is of general application. In plants the chlorophyll-bodies and other chromoplasts are almost certainly to be regarded as differentiations of the cytoplasmic substance."[55] Allowing the symbiotic nature of chloroplasts was only the thin end of the wedge. "The facts," Wilson declared, "point rather to the conclusion that all cell organs arise as differentiated areas in the common structural basis of the cell, and that their morphological character is the outward expression of localized and specific forms of metabolic activity."[56]

The belief that all cytoplasmic structures were differentiations formed by the nucleus was reinforced by the development of the Mendelian chromosome theory of heredity during the early twentieth century. At the same time, the labors of most cell biologists continued to focus on the study of the cell in development and inheritance. The evolution of cells received little attention. Indeed, as biology became ever more specialized around the turn of the century into genetics, ecology, embryology, histology, physiology, anatomy, less and less attention was paid to the potential role of symbiosis in evolution. Moreover, the little literature there was on the topic dealt mainly with separate cases, and it was scattered mainly in European journals, though one could sometimes find brief discussions tucked away in specialist texts. The role of symbiosis in evolution and the possibility that all plant and animal cells were symbiotic complexes continued to be developed by a few individuals in several countries. Among the best known advocates of this view during the first two decades of the twentieth century were two Russian botanists, Andrei Famintsyn and Konstantine Merezhkovskii.

4

Symbiogenesis in Russia

> Above all, a plant, an oak for example, is an animal. An enormous animal in which live parasites or rather symbionts, an infinite multitude of small microscopic green organisms, of the species of unicellular "algae," cyanophyceae.
>
> KONSTANTINE MEREZHKOVSKII, 1920[1]

The suggestion that symbiosis was a primordial characteristic of plants and animals faced serious opposition since it was first proposed in the late nineteenth century. That stable symbiotic associations between phylogentically distinct organisms was a phenomenon of general occurrence flew in the face of the overwhelming belief in individual life struggle and the parasitic nature of microorganisms. The proposal that cells were symbiotic complexes relied on evidence for the relative autonomy of intracellular constituents. And this, in turn, had to be weighed against the arguments that the nucleus alone controlled the synthetic processes of the cell.

Essentially, there were two complementary investigative strategies for demonstrating the role of symbiosis in cell evolution. One involved revealing the morphological and physiological features of specific cellular bodies, identifying these features in specific "free-living" microorganisms, and providing a plausible evolutionary sequence of symbiotic stages that may have led to the origin of plant and animal cells. The other was to experimentally demonstrate the symbiotic nature of the cell by extracting its constituent parts and cultivating them in vitro. During the first two decades of the twentieth century, these two approaches were pursued independently by Konstantine Sergeevich Merezhkovskii (1855–1921) and Andrei Sergeevich Famintsyn (1835–1918). Merezhkovskii adopted the first; Famintsyn, the second.

In recent years, Famintsyn and Merezhkovskii have been adopted as the founding fathers of evolutionary symbiology. Some have stated that they were representative of a tradition of Russian symbiogenesis theorists that remained neglected in the West after being introduced by E. B. Wilson in *The Cell in Development and Heredity.* This view is summarized by Lynn Margulis:

> "Symbiogenesis," the evolutionary origin of new morphologies and physiologies by symbiosis, has been at the forefront of Russian concepts of evolution since the last century. . . . In retrospect, however, these early Russian biologists, introduced to the West by E. B. Wilson (1928), are clearly the founders of "evolutionary symbiology."[2]

However, these claims present a number of difficulties. The suggestion that symbiosis was a source of evolutionary novelty was embedded in the concept of symbiosis from the very beginning. That the cell itself was a symbiosis had been discussed by biologists in Germany, England, and the United States before Famintsyn and Merezhkovskii began to write on the topic. Wilson never mentioned Faminstyn in any of his writings; he mentioned only one of Merezhkovskii's early papers very briefly, and in a way that could hardly be called an introduction.[3] Nonetheless, Merezhkovskii's theory of cell origins was reviewed in detail by the British protozoologist Edward A. Minchin in 1915 in his address as president of the zoology section of the British Association for the Advancement of Science. His essay "The Evolution of the Cell" was republished the next year in *The American Naturalist.*[4] But, more important, Famintsyn and Merezhkovskii did not need Wilson or anyone else to introduce them to a Western audience, for their important papers on symbiosis were published in the West.

Although Khakhina and Margulis recently emphasized that Boris Michailovich Kozo-Polyanski (1890–1957) had suggested, in the 1920s, that cell motility originated by symbiosis,[5] it is doubtful that symbiogenesis was at the forefront of Russian evolutionary thought. During the early twentieth century, Famintsyn and Merezhkovskii themselves made statements actually lamenting that symbiosis as a means for synthesizing new organisms was not taken into consideration in discussions of evolutionary theory in Russia. Indeed, such ideas are completely absent from Daniel Todes's book *Darwin without Malthus,* which focused on issues of cooperation in late nineteenth-century Russian evolutionary biology; they are not mentioned in Mark Adams's writings about evolutionary biology in Russia;[6] and Alexander Vucinich's *Darwin in Russian Thought* notices symbiosis only in passing when discussing Famintsyn's attitudes toward Darwinism.[7] However, Todes does argue that mutual aid had become a common feature of Russian evolutionary thought following Kessler's address on mutual aid at St. Petersburg University, where Kropotkin had first learned of it and where Famintsyn himself was professor of botany.

The Experimental Ideal

Famintsyn was a prominent figure in Russian biology. He was founder of Russia's first laboratory of plant physiology at the Academy of Sciences, and he was well-known among botanists, both inside and outside Russia, for his investigations of various kinds of symbioses involving algae in animals.[8] These studies, especially those on lichen, encouraged him to view mutualistic symbiosis as a mechanism of evolutionary progress from the simple to the complex.

Studies of the dual nature of lichens were carried out in Russia from the late 1860s. As elsewhere, there were conflicting beliefs about the nature of the relationship between the component parts. The Russian lichenologist Alexandr Elenkin was well-known in Europe for his support of the parasitic master–slave interpretation.[9] In his view, lichens were not autonomous organisms; they had to be classified as parasitic fungi, as Schwendener had suggested. Famintsyn opposed this interpretation, and, following Reinke and Frank, he maintained that a mutualistic relationship between alga and fungus had led to the construction of a new morphological whole. As an evolutionist, Famintsyn is known today for his advocacy of a 'psychological approach' to evolution and his neovitalist claims that life could not be reduced to physicochemical analysis.[10] Vucinich has argued that Famintsyn's principal critique of Darwinism was that it did not recognize that organisms played an "active" role in their own evolution and that he advocated a "mental" factor as a key evolutionary element even among "the lowest forms of life."[11]

Famintsyn's first paper on the symbiotic nature of the cell was reported to the Imperial Academy of Science of St. Petersburg in 1906 and later in 1907 in the German journal *Biologisches Centralblatt:* "Symbiosis as Means for the Synthesis of Organisms." Famintsyn began with the assertion that studies of symbiosis offered experimental proof of evolution. Though de Bary had made a similar argument twenty-five years earlier, Famintsyn could still argue that symbiosis had not been taken into consideration in discussions of evolution.[12] He offered an account of the development of his own ideas concerning lichens and suggested that all organisms had similarly emerged as "consortiums."[13]

He stated that since 1868, after being convinced of Schwendener's demonstration of the dual nature of lichens, he had attempted to extract chloroplasts out of plants and culture them. He saw them as corresponding to the "gonidia" (algal symbiont) of the lichen.[14] He had not been successful. In the meantime, cytological work supported his view that "symbiosis was one of the means of the synthesis of organisms." "It turned out that the nuclei of cells, chromatophores and several other components of the cell were not formed bit by bit out of plasma by differentiation, as had been previously assumed." It had been "proven" that each one of these components of the cell lived a "totally independent life and reproduced its kind exclusively by division."[15] In the late 1880s and early 1890s, Famintsyn made his own experimental attempts to extract and culture zoochlorella and zooxanthella from lower animals with the idea that if he were not successful, he would give up hope of culturing chromatophores from plant cells. His attempts were successful.[16]

There was still another reason why Famintsyn published his views on the symbiotic nature of the cell in 1906 and 1907. Merezhkovskii had published his first paper on the symbiotic nature of chromatophores the year before. Famintsyn was anxious to protect what he saw as his priority. At that time, Merezhkovskii was only Privatdozent at the University of Kazan and Faminstyn was quick to dismiss his authority. He remarked that while reading Merezhkovskii's work, "I could not help suspecting that the author knows little about this question and in addition possesses only vague knowledge of the literature that deals with this subject."[17]

Famintsyn repeatedly (and erroneously) claimed that he was the first to consider the evolutionary implications of symbiosis, as exemplified in lichen symbiosis. Some neo-Darwinians, including August Weismann, agreed with him that lichens represented proof of evolution. But Weismann did not consider symbiosis to be a widespread phenomenon, a mechanism that would have to be incorporated into any theory of evolution. The case of lichens was treated as an exception.[18]

Famintsyn concluded his paper by laying out a program for future work. It entailed searching for and studying different kinds of symbiosis in more highly developed forms; examining the constituent parts of the plant cell—chromatophores and colorless plasma and nucleus—separately; and further dividing these parts into smaller organisms. If this work was successful, the next step would be to try to reconstruct the cell synthetically by combining the colorless and green-colored organisms, as had been done with lichens.[19] He continued his efforts to extract these parts and culture them. He reported some results along these lines in 1912. His experiments were on two kinds of freshwater algae: vaucheria taken from the Neva River, and bryopsis, which he studied during a six-month stay at the Muséum Océanographique de Monaco in 1909.[20] Since the 1880s several German botanists had reported the following phenomenon: when the stem of vaucheria is cut in water and some of the substance in the stem of the plant leaks out, spherical bodies are formed which subsequently grow into new vaucheria. Famintsyn repeated these experiments and confirmed previous reports that these amoeba-shaped bodies grow and divide and are transmitted to successive generations through the spores of the host plant. He compared them to mycorrhizal fungi living in the roots of orchids and to zoochlorella living in hydra.[21]

Famintsyn's experiments on *Bryopsis muscosa* were mainly concerned with observing the development and movement of chlorophyll bodies. Through a complicated procedure, he claimed to have observed one chlorophyll grain "change shape and turn into a spherical body similar to a Zoochlorella."[22] He concluded his paper of 1912 with the proclamation that it represented "the first attempt in botany to deal experimentally with the question" of whether the plant cell was an indivisible unit of life or whether it was a symbiotic complex of two or more organisms. He recognized that his results were hardly conclusive proof, but all he could do was repeat what he had stated six years earlier and offer guidelines for future studies.[23]

Famintsyn associated his claims with results of zoologists, especially in the United States, who cultured organs, tissues, and cells: an extracted human heart flexed for twenty hours, a heart of a monkey for fifty-four, a heart of a rabbit for five days. Isolated red blood cells and spermatozoa had been found to stay alive for a considerable time even after long freezing. Some of the most crucial results "in the preservation of life," he reported, were those of the American embryologist Ross Harrison, who managed to culture ganglion cells.[24] Yet there were few botanical attempts to grow cells in culture, and none of the problem of culturing morphologically different parts of the cell.[25]

Famintsyn's call for further experimental work on cultivating parts of the plant cell fell largely on deaf ears. To most cell biologists, efforts to show that the cell was a symbiosis by cultivating parts in artificial media were ill-conceived and doomed to failure. This was true even for the few who, like himself, believed the cell was a symbiosis of elementary organisms. For them, the symbiotic origin of the cell took place in the remote past and the organisms making up the cell were now so well adapted to, transformed by, and integrated into their intracellular environment that it was inconceivable that one could cultivate them outside the cell. This view was shared by Merezhkovskii.

Merezhkovskii Claims Priority

Between 1905 and 1918, Merezhkovskii wrote a series of papers arguing that chloroplasts (chromatophores) were symbiotic microorganisms and that the nucleus and cytoplasm also emerged through a symbiosis of two phylogenetically distinct organisms. In 1910, he offered the term *symbiogenesis* for "the origin of organisms by the combination or by the association of two or several beings which enter into symbiosis."[26] Merezhkovskii's name became so associated with the view that chromatophores were symbionts that during the first decades of the century he was often credited as the originator of the theory. He had done his best to promote this belief, claiming that he was the first to announce this theory in 1905 and that the idea came to him "in a completely spontaneous way" after reading Schimper's classic paper of 1885.[27]

Yet Merezhkovskii had read *The Cell in Development and Inheritance,* where Wilson had discussed the possible symbiotic nature of chloroplasts as well as the ideas of Watasé that the cytoplasm and nucleus (and perhaps centrioles) were symbionts. Indeed, far from the stories of how Merezhkovskii's ideas were introduced to the West, it was more likely the other way around. In fact, Merezhkovskii introduced his first paper of 1905, "On the Nature and Origin of the Chromatophores in the Vegetable Kingdom," with arguments against Wilson's statement that chloroplasts were differentiations of the cell protoplasm.[28] Merezhkovskii did, however, try to move the symbiosis hypothesis beyond the self-reproduction of chloroplasts and their functional analogy with well-known examples of symbiosis of algae in animals.

There were other problems with the proposal that chloroplasts were symbiotic organisms besides demonstrations of their morphological and physiological independence. One had to identify what these symbiotic microorganisms were. Although symbiotic algae were functionally analogous to chloroplasts, algae were known to be very different structurally from simple chloroplasts; they possessed a nucleus and all the classical features of a complete cell. There also seemed to be a logical flaw in identifying chloroplasts with algae per se. For, as the British biologist E. Ray Lankester had asked in 1891, if algae could make their own chloroplasts, why could not higher plants as well?[29] A symbiotic theory for the origin of chloroplasts of "higher plants" would therefore have to account for the origin of the chlorophyll-containing bodies of algae as well. This raised the question of whether there existed simpler "algae" possessing no nucleus and cellular structure as such, and whether some of the algae living in animals such as hydra actually possessed nuclei.

Merezhkovskii attempted to resolve this issue by identifying chromatophores with smaller "algae," cyanophyceae, which he believed lacked a nucleus. The structure of cyanophyceae was one of the most important cornerstones of his theory. He supported his views with a brief comment from the great German evolutionary theorist Ernst Haeckel, who, when discussing the phylogeny of plants, remarked that one could "suppose that cyanophyceae are only individual chloroplasts of real plants." Haeckel had said just enough; not too much so as to undermine Merezhkovskii's own claims to priority: "But, as one can see, it is only a simple assertion, a pure supposition that Haeckel has not even tried to support with any fact, or argument. As a theory, I was the first to establish it in 1905."[30]

Haeckel was well known for his speculative excursions; and his views on the simplest organisms were widely discussed by cell theorists. Since the 1860s, Haeckel had investigated the simplest microorganisms with the expectation that they would give biologists fresh ideas about the origin of life upon the earth, discoveries that would fill the gap between living and lifeless substance and would make the great evolutionary series in the universe entirely uniform. According to Haeckel, the simplest elementary organisms were not cells but organisms which he called "cytodes."[31] These were supposed to be living beings which consisted entirely of an albuminous substance that was not yet differentiated into nucleus and cytoplasm but possessed the properties of both combined. Organisms supposed to be of the nature of cytodes constituted Haeckel's systematic division Monera, of which there was supposed to be two kinds: the Phytomonera and the Zoomonera. The Phytomonera were supposed to have their formative substance colored green and to live in a plantlike manner (cyanophycae). The Zoomonera were colorless amoeboid masses which nourished themselves in the manner of animals. He even claimed to have discovered examples of these, naming them "protoamoeba" and "protogenes," but they had never been found again by any other naturalist.

The idea of structureless organisms, without nuclei, filling the gap between the living and the nonliving, caught biologists' imagination during the middle of the nineteenth century. The first apparent confirmation of the Monera hypothesis came from T. H. Huxley.[32] Sticky mud dredged in 1857 from the North Atlantic depths during the *Challenger* expedition and brought to England for study was seen by Huxley to contain certain lumps of gelatinous substance within which were found numerous granules, but no nuclei. Huxley created a new species and genus of Monera, baptizing it *Baythibius haeckelii,* and believed it to cover large areas of the ocean floor with a layer of primordial protoplasm. Unfortunately, this creature, or rather creation, was often denounced during the 1870s. Huxley himself conceded in 1879 that he had been in error, and he agreed that *Bathybius* was probably a precipitate induced by the application of chemical preservatives (alcohol) to organic substances (gypsum) in the seawater. Many of the other organisms Haeckel classified as Monera had subsequently been found with nuclei. Nonetheless, some leading cell biologists continued to insist that simple cells made up of protoplasm lacking a nucleus had existed in the remote past, if not today.[33]

Mycoplasm and Amoeboplasm

The existence of Monera was central to Merezhkovskii's symbiotic theory of the origin of plant and animal cells. In 1910, he elaborated his arguments that the nucleus of the cells of plants and animals also represented a colony of bacteria living in symbiosis with another organism made up of simple protoplasm.[34] That the nucleus and cytoplasm represented symbiosis between two kinds of organisms had been argued by Watasé in 1893. The idea reemerged in the writings of the acclaimed German cytologist Theodor Boveri. In 1904, he suggested that the nucleated cell arose from a "symbiosis of two kinds of simple plasma-structures—Monera, if we may so call them—in a fashion that a number of smaller forms, the chromosomes, established themselves within a larger one which we now call the cytosome."[35] However, Boveri did not develop this idea. Instead, he is best known, not as a theorist, but as brilliant cytologist and an ingenious experimentalist, one of the architects of experimental embryology.

Merezhkovskii did develop it. He not only speculated on the physiological properties of the organisms involved, he also offered an evolutionary account—a reconstruction of the origin of plant and animal cells from two sorts of protoplasm, which he called *mycoplasm* and *amoeboplasm.* Each was supposed to differ fundamentally in nature and to have historically distinct origins in different epochs of the earth's history. Bacteria, the chromatin grains of the nucleus, and chloroplasts were held to be of the nature of mycoplasm, the cytoplasm of amoeboplasm.

Merezhkovskii divided the history of the earth into four epochs. In the *first,* the earth was an incandescent mass of vapor; in the *second,* it had a firm crust. Mycoplasm was the first type of protoplasm which came into existence during what he called the *third* epoch, at a time when the crust of the earth had cooled sufficiently for water to be condensed upon it, but when the temperature of the water was still near the boiling point. Consequently, the waters of the globe were free from oxygen while saturated with all kinds of mineral substances. The origin of the second type of protoplasm, amoeboplasm, was supposed to have taken place during a *fourth* terrestrial epoch when the waters covering the globe were cooled down to below 50°C and contained dissolved oxygen but few mineral substances. The different nature and constitution of these two types of protoplasm corresponded with the differences of the epochs and the conditions under which they arose.

The earliest forms of life were "biococci." These were ultramicroscopic particles of mycoplasm, without organization, capable of existing in temperatures close to the boiling point and in the absence of oxygen. Biococci possessed the power to build proteins and carbohydrates from inorganic materials, and they were very resistant to strong mineral salts and acids and to various poisons. Bacteria arose from biococci and were for a long time the only living inhabitants of the Earth. Later, when the temperature of the terrestrial waters had been lowered below 50°C and contained abundant organic food (bacteria), amoeboplasm emerged in small masses as nonnucleated Monera. They crept in an amoeboid manner on the floor of the ocean, devouring bacteria. The next step in evolution occurred when, in some cases, bacteria ingested by the nonnucleated Monera resisted digestion and maintained a symbiotic existence in the amoeboplasm. At first, the symbiotic bacteria were scattered in the Moneran body, but they later became concentrated at one spot, surrounded by a membrane, and gave rise to the cell nucleus. This was an immense step forward in evolution, since the locomotor powers of the simple and delicate Monera were now supplemented by the great capability possessed by the bacteria of producing ferments of the most varied kinds.

Meanwhile, the free bacteria continued their natural evolution and gave rise to the cyanophyceae and to the whole group of fungi. The plant cell came into existence by a further process of symbiogenesis, in that some of the cyanophyceae, red, brown, or green in color, became symbiotic in nucleated cells, for the most part flagellates, in which they established themselves as the chromatophores, or chlorophyll corpuscles. This is how Merezhkovskii believed that plant cells originated and thus the evolution of the Vegetable Kingdom started as a double process of symbiosis. Those amoeboid or flagellated organisms which formed no symbiosis with cyanophyceae continued to live as animals and started the evolution of the Animal Kingdom.

As a logical deduction from his theory, Merezhkovskii classified organisms generally into three groups or kingdoms: the Mycoidea, comprising bacteria, cyanophyceae, and fungi, and in which no symbiosis had taken place; the Animal Kingdom, in which true cells have arisen by a simple symbiosis of mycoplasm (chromatin) and amoeboplasm (cytoplasm); and the Vegetable Kingdom, in which nucleated cells had entered upon an additional symbiosis with cyanophyceae.

Refuge in Geneva

Merezhkovskii's first papers on symbiogenesis were published in German as well as in Russian. However, during the turbulent years of World War I he had taken refuge in Geneva. It was there, two weeks before the the end of the war, on 25 October 1918, at age seventy-three, that Merezhkovskii gave his last paper on what he considered to be "the work of his life." This presentation, entitled "La plante considérée comme un complexe symbiotique," was published two years later as a lengthy paper in *Bulletin de la Société des Sciences Naturelles de l'Ouest de la France.* Merezhkovskii was bitter, sick, and tired. As he lamented, since announcing his theory on the symbiotic nature of chromatophores in 1905, "it had made little headway"; it was often "completely ignored." He had the intention of writing a book on the subject, but other work got in the way; then, during the political struggle in Russia before the war, he became too weak to write it; "finally the war, the revolution. . . ." Thus he remarked, "It is only today, on the eve of leaving this sad world, that I have decided to develop my ideas in a little more detail in consolidating and enlarging the basis on which they rest."[36]

While in Geneva Merezhkovskii had approached several French botanists for help in providing a forum for him to speak on symbiogenesis. But he found no support. In fact, the professor of botany at the University of Geneva, R. Chodat, went out of his way to prevent Merezhkovskii from speaking. Merezhkovskii published the details.[37] Two zoologists, Yves Delage of the Académie des Sciences, Paris, and A. Labbé, president of the Société des Sciences Naturelles de l'Ouest de la France, came to his aid. Merezhkovskii declared, "One day the history of botany will thank them."[38]

Professor of zoology, anatomy, and comparative physiology at the Sorbonne and director of the marine biological station at Roscoff, Yves Delage (1854–1920) was one of the old savants of French biology. He had written one of the great texts reviewing and criticizing nineteenth-century theories on the cell, heredity, and variation: *La Structure du Protoplasma et les Théories sur l'Hérédité et les Grandes Problèmes de la Biologie Générale* (1895). His book with E. Hérouard, *La Cellule et les Protozoaires* (1896), was one of the few protozoology texts of its time.[39] Together with the anarchist neo-Lamarckian biologist Marie Goldsmith, Delage edited the journal *L'Année Biologique,* which kept biologists abreast of recent contributions, reviewing them with their own commentary. One of the early neo-Lamarckians in France, Delage was always on the lookout for alternative theories to neo-Darwinism, and he maintained a longstanding interest in symbiotic theories of the cell. In fact, a great deal of controversy about the role of symbiosis in heredity, development, and evolution was brewing in France at the very time Merezhkovskii was speaking.

The Sacred Fire of the Lighthouse

At Geneva, a few weeks before the end of World War I, Merezhkovskii laid out carefully what he regarded as the proofs that chromatophores represented organisms introduced into the cell from outside. He rested the edifice of his theory on three pillars: (1) the continuity of the chromatophores; (2) the structure of cyanophyceae; and (3) the comparative structure and comparative physiology of chromatophores and cyanophyceae.[40] Each was controversial, and Merezhkovskii treated them in turn. The general opinion propagated in all the botanical treatises was that the chromatophores appear in the interior of plant cells as a result of the gradual differentiation of the protoplasm of the cytoplasm. Yet, he argued, chromatophores do not appear anew; they reproduce by division and are not included in the hereditary substance of the nucleus. He asserted that in 1906 he himself had shown this to be true for the unicellular algae Diatomes, but that Schimper and his many successors had provided indisputable demonstrations that chromatophores of plants exist in the form of colorless leucoplasts in the egg, or in the form of chloroplasts in the spores of a plant. When the eggs or spores of a plant divide to form tissues of the new plant, the chromatophores also divide and distribute themselves in the new cells. Thus there was an uninterrupted continuity of chromatophores. They were never formed anew.[41]

That chromatophores were self-reproducing, in Merezhkovskii's view, proved that they had been independent organisms which entered into the interior of cells. If they were originally part of the cell plasma at the beginning of the Vegetable Kingdom and later acquired the ability to reproduce themselves and be transmitted from one generation to the next, one would have to explain why the plasma had lost the ability to make chromatophores.

Further, he argued, the theory that chromatophores emerged by differentiation concealed "a gross mistake in logic and lacked any comprehension of the essence of heredity."[42] The chromatophore was a complicated body comprising at least four different parts: clear uncolored plasma (stroma), green granules, the pyrenoid, and finally the different coloring materials of chlorophyll. Had a body so complicated appeared by way of differentiation out of cell plasma, it could have emerged only gradually in a step-by-step fashion. Each stage in the evolution of the chromatophore would then have to be incorporated into the hereditary constitution of the nuclear chromosomes. But if this had occurred then, the chromatophore would have to be formed anew in the egg in each generation under the influence of the chromosomes. But this was contrary to the facts. Therefore, Merezhkovskii reasoned, "The continuity of chromatophores proves that they are organisms which came from outside and by consequence that the plant is a symbiosis."[43]

But if all this was so apparent, so obvious, so logical, then why had not others—the botanists in Germany who had shown the continuity of the chromatophore—reached the same conclusion? Merezhkovskii claimed that Schimper himself had overlooked this profound insight, and he answered this query by appealing to anti-German sentiment. It could be explained by the "particular properties of the German mind." Germans were strong in the accumulation of facts, but weak in generalizing and constructing theories on these bases:

> The Germans compare German science to a lighthouse. I would as well, but then to a lighthouse without sacred fire to light up the world. The Germans carry stones to construct a solid base without which there would be no lighthouse, and in this task no other nation surpasses them. But, it is left for others to arrive and light the fire. Now, without fire, there is no lighthouse.[44]

This was not simply empty rhetoric to please an audience in the midst of a war with Germany; it played an important role in Merezhkovskii's argument for relying so much on the work of German botanists and yet protecting his own originality as a theorist. That chloroplasts represented independent symbiotic organisms had been discussed by Reinke, Schimper, and Haberlandt in Germany, Lankester in England, Watasé and Schneider in the United States, and several others; Merezhkovskii never referred to any of them. An oversight, due to lack of the relevant journals in Russia? Perhaps; but unlikely. Not only had he read E. B. Wilson's book *The Cell,* where these views had been discussed, but he had also cited the paper in which Theodor Boveri mentioned the possibility that the nucleus and cytoplasm were symbionts.[45] Yet he never referred to Boveri's views on the matter. In his lengthy paper of 1918, he also never referred to Famintsyn's discussions of the idea that chloroplasts were symbionts.[46] Surely Merezhkovskii could have made good use of such a formidable list of allies. Instead, he painted himself into a corner, portraying himself as an oppressed individual with only the naked truth on his side, struggling against an unjust and uncaring scientific community.

These remarks should not detract froom Merezhkovskii's originality, for no one had developed the argument for the symbiotic theory of the cell to the extent that he did. Famintsyn notwithstanding, the few biologists who had read Merezhkovskii's work, though critical, regarded his schemes to be ingenious. He was able to unite a wide body of apparently disparate scientific results and reach astute conclusions. Merezhkovskii, however, was not quite as generous in his appreciation of others. He belittled the theoretical abilities not only of those Germans who investigated the nature of chromatophores but also of those who investigated the structure of cyanophyceae, the second pillar upon which he rested his theory.

Since the 1880s the structure of cyanophyceae had been the subject of numerous studies. Generally they were understood to be a group of plants which resembled the algae in some respects but were sharply differentiated from them in the structure of their cell contents. Observation of these extremely small organisms with existing microscopes was far from straightforward, and several different interpretations of their structure and phylogenetic relationship to other groups were offered. The central question concerned the presence or absence of a nucleus and/or chromatophore.

Merezhkovskii reduced the debate to two competing camps in Germany led by Kohl and Fischer. According to Kohl, cyanophyceae were comprised of numerous granules or chromatophores surrounding a *nucleus* with chromosomes. Fischer, on the other hand, argued that they were comprised of one chromatophore, a hollow sphere surrounding *cytoplasm* at its core. In Merezhkovskii's view, cyanophyceae possessed neither nucleus nor chromatophore; the cyanophyte as a whole was a single chromatophore. The central body was neither a nucleus nor cytoplasm; it represented a pyrenoid, the starch-forming center of a chromatophore. Fischer, he argued, had almost put his finger on the truth but still could not see that the cyanophyte itself was a chromatophore. Merezhkovskii commented that here again one could see a clear expression of the particularities of the German mind and German science. Fischer had gathered the rocks necessary for constructing the lighthouse, executing his work with a knowledge and skill that Merezhkovskii could never attain. "This said, he stopped himself: the forces are lacking in him to raise himself above the facts. And as it was also in the case of Schimper, it has been left to another to arrive and light the sacred fire of the lighthouse."[47]

The only difference between the cyanophyte and the chromatophore was that the former had membrane composed of a nitrogenous substance whereas the chromatophores lacked a membrane. They also shared two remarkable physiological properties. Both were able to transform inorganic substances into carbohydrates, and to produce the synthesis of albuminous substances out of inorganic substances. The well-established evidence of the symbiosis involving cyanophyceae also lent support to the claim that chromatophores of plants were in reality cyanophyceae. *Anabaena cycadeae,* living in the intercellular spaces of the roots of cycads, highlighted by the British botanist Ethel Rose Spratt, merited special attention because this showed the great tendency of cyanophyceae to enter into symbiotic relations with various kinds of organisms.[48]

Merezhkovskii further supported his claim of the symbiotic origin of chromatophores with algae (zoochlorellae and zooxanthellae) living symbiotically in protozoa, freshwater sponges, hydra, and certain turbellarian worms. By 1918, he could argue that symbiotic algae had been found in almost every class of "lower invertebrate." Protozoa containing cyanophyceae were actual examples of "animals on the way to becoming plants." "In creating these strange beings," these "plant-animals," he argued, "nature seems to have wanted to give us a demonstration *ad oculos* of the symbiotic nature of the plant and teach us how the vegetable world had taken birth."[49]

The phylogeny of lichens offered Merezhkovskii a model for constructing phylogenetic trees of plants. Lichens had a polyphyletic origin of the highest degree: each of the (at least ten) phylogenetic lines of this order had a completely independent origin from the others, and each had originated by a combination of different fungi with different algae and diverse cyanophyceae.[50] Similarly, he argued that plants had multiple origins due to symbiosis. For Merezhkovskii, all algae derived from symbiotic combinations of diverse flagellates with diverse cyanophyceae. Brown algae originated from the combination of a flagellate and a brown cyanophyte, red algae from a combination of a flagellate with a red cyanophyte; green algae from ciliated and flagellated infusoria.[51] As Merezhkovskii remarked, "I believe it would be no exaggeration, to say that the vegetable kingdom had originated at least fifteen times."[52]

In the phylogenetic landscape he imagined, both lichens and plants "represented a grove of immense oak trees isolated on a meadow." The Animal Kingdom, represented mainly by one single branching tree of Metazoa, stood in stark contrast to these groves. But the animal world was also the result of a symbiosis—that of different Monera (cytoplasm) with different Bacteria (nucleus). Growing around the large tree of Metazoa were small groups of Sporozoa of independent origin which had acquired the dimensions of phylogenetic "shrubs." In addition to these, there were very small groups of infusoria of independent origin which represented "the grass of the lawn on which the great tree of animals grows."[53]

Merezhkovskii did not discuss his theory of symbiogenesis in relation to the struggle for existence. With regard to mechanism, in the sense of cause and effect, symbiogenesis as he conceived it was a teleological concept. He was a mystic, as was his brother Dimitri Sergeevich, a famous religious philosopher who organized a religious philosophical society in St. Petersburg during the first decades of the century. Merezhkovskii considered symbiogenesis and the origin of the plant world from the point of view of transcendental philosophy. He supposed that the ultimate goal of the universe was its "complete spiritualization, that is, the gradual transformation of all the mechanical energy of the Universe into psychic energy." In creating the vegetable world nature followed the principle of the division of labor:

> The vegetable world seizes solar energy and transforms it into potential chemical energy in the form of albuminous substances, so that the animal world, in nourishing itself on plants, can with so much more amplification transform it into psychic energy. The spiritualization of the Universe, far from losing by this division of labor which exists between the two organic worlds, only gains by it.[54]

The interpretation of symbiosis based on a division of labor, exemplified by reciprocal relations between plants and animals, was developed by other biologists during the first two decades of the century. And when constructed in terms of cooperation for the common good, the concept of symbiosis continued to oppose an amoral view of cosmic processes in which nonhuman organismic interaction was understood in terms of a multitude of selfish interests. This opposition became more evident during World War I, when "the struggle for existence," interpreted in terms of the war of nature, was used by some to sanction militarism.

5

Engendering Genesis Stories

> Thus arose in the beginning the brand of Cain, the prototype of the animal, that is to say, a class of organism, which was no longer able to build up its substance from inorganic materials in the former peaceful manner, but which nourished itself by capturing, devouring and digesting other living organisms.
>
> E. A. MINCHIN, 1915[1]

When the Great War began, there were attempts to explain away the atrocities in terms of "the struggle for existence." A protest against such an "abuse" of Darwin's terminology appeared in a letter published in the London *Times*. It was said that such an explanation was "little more than an application to philosophy and politics of ideas taken from crude popular misconceptions of the Darwinian theory (of 'struggle for existence', and 'will to power', 'survival of the fittest' and 'superman,' etc.)"; and that there was, however, a work in English "which interprets biological and social progress not in terms of overbearing brute force and cunning, but in terms of mutual cooperation."[2] A reprint of Kropotkin's *Mutual Aid* was quickly published to emphasize that the war was not the result of an inherent aggression among the masses. It was their rulers who prepared the war, and their intellectual leaders who worked out its barbarous methods.[3]

Mutual Aid was not the only book published in England during World War I that opposed the interpretation of biological and social progress in terms of the "struggle for existence." In 1915 Hermann Reinheimer's *Symbiogenesis: The Universal Law of Progressive Evolution* appeared, bearing a similar message.[4] Symbiosis, understood in terms of mutual aid, not individual life struggle, was the means of progressive evolution. Reinheimer is virtually unknown among contemporary biologists and historians. He wrote several books, including *Nutrition and Evolution* and *Survival and Reproduction*. His most popular and well-received book was *Evolution by Co-operation: A Study in Bio-economics* (1913).[5] *Symbiogenesis,* which he wrote over eighteen months of "almost uninterrupted daily labour" and with the help of the well-known biologist A. H. Singleton, followed directly from it.[6]

All of Reinheimer's arguments were directed against the claim that evolution had no morality and that it was due solely to the blind agency of natural selection. *Symbiogenesis* owed little if anything to Merezhkovskii, to whom Reinheimer never referred. On his own account, he was following in the tradition of late nineteenth-century mutual aid theorists such as Kropotkin as well as Henry Drummond's writings on *The Ascent of Man* in which he maintained that the struggle for the life of others was more important than the struggle for individual life.[7]

Feminine Nature

Reinheimer's views embraced the criticisms of many biologists who, during the decades around the turn of the century, voiced opposition to Darwinism. Indeed, this period is marked by what Peter Bowler, following Julian Huxley, has called the "eclipse of Darwinism."[8] Various alternative theories which downplayed the role of natural selection were emphasized. Most prominent among them was a renewed emphasis on the inheritance of acquired characteristics. Reinheimer saw his theory of symbiogenesis to be in line with those neo-Lamarckian views. From this perspective, life was not simply a matter of increased rates of fecundity, and evolution would not act through random changes in structure. On the contrary, new functions led to new structure. The evolution of new traits would occur through labor—through the use and disuse of parts. Neo-Lamarckian biologists generally attributed a more active role to organisms in their own evolution; they were not simply responding passively to external conditions and selection pressure.

An element of Lamarckism was present in the original version of Darwinism, and Darwin himself, toward the end of his career, emphasized more and more the significance of use and disuse on heredity. Spencer and Kropotkin had also supported a combination of selection and Lamarckism. In Germany, Haeckel had allowed Lamarckism an even more prominent role. By the turn of the century many neo-Lamarckian naturalists adopted the theory of orthogenesis, or progressive variation along a definite line, as associated most prominently with the German Theodor Eimer's *Organic Evolution,* which was translated into English in 1890.[9]

Eimer adopted an organicist concept in which the diversity of individuals in nature was understood as a division of labor functioning for the common good. He denied that the "individual," in the sense of an indivisible organic unit, existed, since all single forms were directly or indirectly connected with one another. One could only "artificially" establish individuals by defining one's conception in each case in a particular way.[10] Eimer interpreted his organicism as entailing the notion that if the individual did not struggle for the good of all, if would only do injury to itself.[11] Taking as a model the social life of bees, in which the work of the individual for the community had become automatic action, he argued that the morality of working for the common good had slowly evolved as an instinct among humans.[12]

Reinheimer also allied his views with the moral philosopher Henri Bergson,[13] who argued that theory of knowledge and theory of life were inseparable and that there could not have been a break between prehuman nature and "ethical man," as T. H. Huxley had asserted: "If natural selection is the law of nature, it stands to reason that it must assert itself in the end over all man-made laws."[14] Reinheimer agreed with those who had asserted that "Darwinian teaching has had some share in bringing about the present plight of Europe."[15] His aim was to replace natural selection with the cooperative "laws of symbiogenesis," whereby all organisms, including humans, were equal in that they were "responsible members of the web of life."[16]

Reinheimer declared that from the moment division of labor set in, cooperation had been essential to the organic world, for it afforded the best available means of preservation and progress. He stated that it was the bioeconomic task of organisms to earn their sustenance, and over and above this to provide for marginal and exchangeable "bioeconomic values" to be used in the mutual accomplishment of evolution: "Just as in human society an invisible but nevertheless actual and indispensable standard of useful morality has gradually arisen, so also the biological world is ruled by an invisible but actual standard of domestic and bio-economic usefulness or social morality."[17]

Reinheimer dismissed the whole edifice of Darwinism as being based on a series of loose metaphors and semitheological verbiage.[18] It represented a pseudo-economic outlook on life borrowed from the "most sterile portion of political economy, represented by the Malthusian doctrine." Darwinism provided a distorted and pessimistic picture of the relations of organism to the universe, whereby "their very presence adds to the difficulties of organic existence."[19] He insisted that the Malthusian assertion that there is a constant tendency in all inanimate life to increase beyond the nourishment prepared for it did not make adequate allowance for the share which the organism itself bears in the production of sustenance or of its equivalents. It also failed to recognize that organisms themselves practiced restraint in reproduction in response to variations in their environment.[20] He asserted that to bring in economics only at a point where an excessive number of young were fighting each other for an insufficient supply of food was to establish an arbitrary system of bioeconomics. Darwinism, he contended, began at an abnormal point where wastefulness has led organisms into a cul-de-sac, successfully avoided by great classes of others.[21]

Reinheimer's account began with nutrition as a dominant factor in the evolutionary process. Healthy organisms and thriving species engage in symbiotic reciprocity, comparable to trading individuals and communities, whose claim to survival rests on the production of commodities or the rendering of services indispensable to others. In the course of the development of this "civilization" there was an "accumulation of values of food reserves or surplus." Labor was incessantly stored in the form of biological capital (heredity), and there was a continuous exchange of surpluses that was as indispensable to the preservation and the progress of the organic world as to that of the body politic.[22]

He advocated the existence of a law relating to nutrition analogous to that stated by Darwin with regard to fertilization. "In-feeding is abhorred by Nature," just as, according to Darwin, perpetual "in-breeding is abhorred by Nature." By the term "in-feeding" he denominated the indolent appropriation of food manufactured by close relatives, and the correlated shirking of the economic duty of production, or of mutual services in some kind. In-feeding was the source of economic sterility and consequent limitation and insecurity of life. By the term "cross-feeding," on the other hand, he designated what he conceived to be the norm of healthy nutrition based on symbiotic reciprocity. No new bioeconomic values were created by parasitism, which was based on amoral relations of in-feeding. Mutual reciprocity was the normal and the progressive; selfish interests and parasitism represented the pathological and the degenerate.

Mutual relations between plant and animal kingdoms were fundamental for progressive evolution. "Bio-economically speaking," Reinheimer asserted, it was "the duty of the plant world to manufacture the food-stuffs for its complement, the animal world. In so doing it not only ensures its own preservation and that of the animal, but improves its own security and status together with those of the animal."[23] The majority of the foods of the world are products of plant reproduction—"love foods," as Henry Drummond had called them.

Reinheimer referred to the "Eternal Feminine in Nature," of Mother Earth, Gaia: "It is the meek, anabolic, vegetative and female element in particular that is characterized by survival value."[24] He asserted that the general "educative" influence of the plant kingdom upon that of the animal during evolution was comparable to that which the "female of the species" has had upon man. The plant kingdom in numerous silent ways managed to "contrive the success of the truly useful, and to bring about the ultimate decline of those organisms which misused the established relation of 'give and take' between the two kingdoms."[25] Wholesome plant food was the means of importing vital "earth-force":

> We speak of "mother" earth, and in Greek mythology Antaeus, the giant of Libya, the son of Poseidon and Gaea, when thrown in combat, derived fresh strength from each successive contact with his mother earth, which is figurative of the way in which we still obtain "Antaeus"-force from mother earth by remaining in adequate contact with her.[26]

Adequate contact with Gaia maintained by proper nutrition (cross-feeding) produced a "nutritive amphimixis analogous to the fusion of reciprocally differentiated germ cells in sexual amphimixis."[27] In *Symbiogenesis,* he argued that the knowledge to be gleaned from studies of symbiosis confirmed these views and filled the lacuna in Darwin's theory.[28]

Reinheimer defined symbiosis in terms of cooperation, but it could not be simply restricted to a physiological partnership between individuals of different species. As Spencer had argued before him, he maintained that all forms of attached cooperative partnerships were foreshadowings or variations of the wider forms of cooperation which underlay evolution and united all organisms in one vast web of life "in a veritable organic civilization."[29] Symbiogenesis referred to "the principle underlying such symbiosis and indeed all instances of mutuality in the progressive transmutation of biological values generally."[30]

Symbiogenesis involved a definite evolutionary path, and pathogenesis an opposite one, which led accordingly to "a well defined biological antagonism between the 'good' (the symbiogenetic) and the 'evil' (the predacious, indolent and pathogenetic) types."[31] Symbiosis through cross-feeding habits was a relation which could not without impunity be violated for any length of time. That the opposite mode, "in-feeding," was as fallacious as perpetual inbreeding, Reinheimer argued, was "the greatest lesson for contemporary humanity to derive from evolutionary studies."[32]

Convoluta

To show the relationship of primordial symbiosis to larger relations of plants and animals, Reinheimer turned to studies of the marine worm *Convoluta roscoffensis* as discussed by Sir Frederick Keeble in his little book *Plant-Animals: A Study in Symbiosis* (1910).[33] These almost microscopic unsegmented organisms, which colored the sand green around Roscoff in Britanny, possessed a well-defined nervous system and efficient sense organs, but a primitive digestive tract, and they were without any proper circulation or excretory apparatus. Soon after they are hatched these creatures swallow with their food certain algal cells. For a time they continue to feed and grow, and as they do the swallowed algal cells divide and multiply, until eventually the masses of these cells occupy a great part of the animal's body and turn it green, and the worms cease to ingest solid food substances.

A similar apparently total abstinence had been reported for certain adult Radiolaria, Ciliates, Hydrocorallines, and Madreporaria. No food remains were present in the adult stages of these animals, and the inference was drawn that such animals subsist on food materials manufactured solely by their green or yellow cells.[34] In various experiments, in light and darkness, Keeble showed that *C. roscoffensis* depend for their food on these algal cells; without them they fail to grow.[35] Keeble conceived of *C. roscoffensis* as an animal that lives like a plant—a plant-animal. The algal cells build up carbohydrates while the animal supplies it with its nitrogen wastes. The animals, looking to the algae to come and take charge of the work of getting rid of these waste substances, have ceased to construct any excretory apparatus whatever.[36]

As for the algal cells of *C. roscoffensis,* they were much simpler than those found in other animals. They did not possess a nucleus and a cellulose cell wall. They possessed little more than a green chloroplast—that is, when they were inside the tissue of the worm. Moreover, once the algae entered the worm, Keeble was unable to isolate and cultivate them. Nonetheless, it was clear that they were algae of exogenous origin, because when he grew the worms in filtered seawater, they remained colorless. He was also able to isolate the free-living stage of the alga from the surface of the egg capsules. Keeble cultured it, and he was able to synthesize *C. roscoffensis* by infecting colorless worms with it. However, the alga in the free-living stage was much more complex than when inside the animal. It possessed flagella, an eye spot, a pyrenoid, and a nucleus as well as a cup-shaped chloroplast. Keeble thought it might be a species of *Chlamydomonas*.

But more important was the fact that "the green cells of the body of *C. roscoffensis* once saw independent days, and that, for those cells, naked and deprived of nuclear material, these independent days are gone, never to recur."[37] Keeble believed that the algal cells of *C. roscoffensis* were "on the road to complete loss of independence." The intimacy of this symbiotic relationship lent support to the suggestion that chloroplasts of higher plants were of symbiotic origin.[38]

In the model Keeble supported, chloroplasts were not decendants of minute nonnucleated cyanophytes, as Merezhkovskii had proposed. Like Haberlandt before him, he suggested that chloroplasts of plants descended from free-living nucleated algal cells that had lost everything except their chloroplasts, just as "an adult *C. roscoffensis* is a complex of two organisms: one, the colourless animal, the other, the chloroplasts—remainders of the original, green, nucleated, algal cells."[39] According to this story, "In return for security and all the comforts of home, the green cells prepared the food for themselves and for their hosts. Submitting itself to the guidance of the animal, the green cell abandoned its nucleus and became reduced to a naked chloroplast."[40]

Nonetheless, Keeble recognized that there was a wide gap between the state of affairs in *C. roscoffensis* and that in the higher green plant. The "higher green plants provide for the future crop of chloroplasts in their descendants by transmitting colourless rudiments of the chloroplasts to their egg-cells."[41] But this was far from the case in *C. roscoffensis*. None of its green cells or any colorless representatives pass into the egg cells. When *C. roscoffensis* matures and produces eggs, the animal changes its mode of nutrition and begins to destroy and digest its algal cells, and it continues to do so until its green tissue disappears. The eggs are supplied with carbohydrates at the expense of the green cells, "so even though it now dies as the consequence of its ill-conceived greediness, the continuance of the species is assured."[42]

If one looked at the relationship from the standpoint of the animal, Keeble remarked, "it is one of obligate parasitism." The plant, for its part, "sacrifices its independence for a life of plenty."[43] The plants flourish in the bodies of these animals because they discover large accumulations of waste nitrogen compounds in short supply in the open sea. Nonetheless, the fact remained that *C. roscoffensis* possessed "self-sown, well-tended, highly productive gardens," and Keeble concluded that "if they could but learn how to bequeath packets of vegetative seed to their descendants, they might lose their animal characteristics altogether and become, *C. roscoffensis,* a green plant."[44]

To Reinheimer, the message to be gleaned from *C. roscoffensis* was clear: the relationship was one of work and nutrition; it was one of symbiosis—the antithesis of disease—as long as it lasted. But *C. roscoffensis* failed in biological symbiosis, and the difference between its relations and that of the "domestic symbiosis in the higher green plants" showed the importance of maintaining symbiogenetic principles for progressive evolution. *C. roscoffensis* could not set up "a permanent and mutually beneficial relation with the algae they exploit or produce any bioeconomic values (i.e. social or communal utilities) in virtue of this exploitation." As Reinheimer saw it, this was "retrograde symbiosis" and had to be distinguished from the progressive symbiosis which he believed lichens depicted.[45]

Lichens, for Reinheimer, represented an argument against the Malthusian principle. They showed how "reproduction itself is normally subservient to bio-economic purposes."[46] Lichens were characterized by slow growth and great length of life. Some botanists had calculated that some lichens were many hundreds of years old and that some species actually outrival in longevity the oldest trees. If evolution were merely a matter of achieving reproduction, Reinheimer remarked, there would be little need for slow growth.[47]

The usefulness of the symbiosis was important for both the alga and fungus since the lichen could live in situations where neither partner could exist alone. But the useful effects of this symbiosis reached far beyond the collective interests of alga and fungi. Lichens were also useful to other organisms since, in disintegrating by mechanical and chemical means the rocks on which they live, they played an important role in soil production. By virtue of their symbolic labors they were "veritable pioneers of organic civilization."[48]

Thus Reinheimer opposed the phenomenon of symbiosis, and the principles of reciprocity which underlay it, to the one-sided emphasis on conflict and competition. But this was not the only obstacle to recognition of cooperation or "the laws of symbiogenetic evolution." As both Reinheimer and Eimer saw it, the rise of laboratory-based biology and experimental studies of organisms in isolation was also doing its share in minimizing the role of organismic interaction in evolution.[49] Yet, if this posed a problem from the point of view of naturalists, the few microscopists who investigated primordial symbiosis and cell origins complained of their own problems of attracting interest in this line of research.

Problems with Protista

Constructing phylogenetic accounts of cell origins, whether symbiogenetic or not, was not a popular pursuit around the turn of the century. The obstacles to popularizing such studies were well discussed in 1915 by Edward A. Minchin (1866–1915) in his address as president of the zoology section of the British Association for the Advancement of Science, "The Evolution of the Cell." Minchin discussed symbiotic theories when offering his own account of cell origins. But, as he saw it, any discussion concerning the cell ranked "in popular estimation as dealing with some abstruse and recondite subject quite remote from ordinary life, and of interest only to biological specialists."[50]

The technical requirements, the use of expensive and delicate optical instruments for studying cells, prevented the participation and further limited the interests of the "educated" classes. Yet even those who possessed the necessary training and equipment showed little interest in cell evolution. Most professional cytologists confined their studies of cells to those found in Metazoa and Metaphyta; they were interested primarily in the cell in development and heredity. The organisms they used, and the problems they addressed, simply did not bring them face to face with more "primitive cells."[51]

The origin of the metazoan cell, Minchin argued, had to be sought among the Protista. The study of the Protista had gained increased attention since the last days of the nineteenth century, primarily because of the discovery of the importance of some of the parasitic forms as invaders of the bodies of humans and animals and causes of disease often of a deadly nature. Minchin himself worked at the Lister Institute of Preventive Medicine, Cheslea, and as professor of zoology at the University of London; he is known today for his studies of trypanosomes. His textbook of 1912, *An Introduction to the Study of Protozoa,*[52] was one of a mass of protozoological works that appeared around that time, emanating especially from the pens of his German colleagues.[53]

In 1903 Robert Koch's Institute for Infectious Diseases in Berlin added a special division for protozoology. A few years later, the Imperial Ministry of Health established two new specialized institutes for parasitological protozoological research: the Institute for Protozoology in Berlin and the Hamburg Institute for Naval and Tropical Diseases. In 1906 the Georg-Speyr House for chemotherapeutical research in Frankfurt was opened; headed by Paul Ehrlich, it specialized in biochemical studies on the agent of syphilis and other protozoan diseases.[54]

Though spurred by medical interests, studies of the Protista also yielded results of the utmost importance for general biology. From an evolutionary perspective, Protista were defined not on the basis of any disease-causing attributes but as unicellular organisms.[55] This was the message of one of Germany's leading cell biologists, Richard Hertwig, in the lead article of the first issue of *Archiv für Protistenkunde* in 1902.[56] Minchin had spent a year in Germany in the early 1890s working with Richard Hertwig and Otto Bütschli, and he maintained their view that Protista were unicellular organisms, and that one could use them to study the origins of the cells of multicellular organisms: "We find in the Protista," he wrote, "every possible condition of structural differentiation and elaboration, from cells as highly organized as those of Metazoa or even, in some cases, much more so, back to types of structure to which the term cell can only be applied by stretching its meaning to the breaking-point."[57]

Organisms without Cells

Minchin confined his address of 1915 to the evolution of the cell as seen in its typical form in the bodies of multicellular organisms starting from the simplest conceivable type of living being then known. But, as he well knew, the claim that one could investigate the evolution of the metazoan cell by studying Protists was in itself a subject of contention. Some biologists refused to admit any homology between the individual protistan organism and a single cell of the many that build up the body of a metazoon. They maintained that the individual protistan organism had to be regarded as homologous with the metazoan individual as a whole. They further denied the concept of the cell state. The development of a chick, frog, or any other egg, it was argued, was not a matter of aggregation of millions of separate individual cells bound together by the division of labor and mutual dependence. Instead, the adult organization was identical in its individuality with that of the egg. The formation of cells was of secondary significance.

One could hear protests against cell theory in many countries. In England, T. H. Huxley had offered one of the first critiques, which became frequently cited in the decades around the turn of the century. As early as 1853, he argued against the "erroneous conception of the organism as a beehive," which he attributed to Matthias Jakob Schleiden and Theodor Schwann. He declared that the cells "are no more the producers of the vital phenomena than the shells scattered along the sea-beach are the instruments by which the gravitative force of the moon acts upon the ocean. Like these, the cells mark only where the vital tides have been, and how they have acted."[58]

In Germany, the botanist Julius Sachs made similar remarks in 1887:

> To many, the cell is always an independent living being, which sometimes exists for itself alone, and sometimes becomes joined with others—millions of its like, in order to form a cell-colony, or, as Haeckel has named it for the plant particularly, a cell republic. To others again, to whom the author of this book also belongs, cell-formation is a phenomenon very general, it is true, in organic life, but still only of *secondary significance.*[59]

In the United States, C. O. Whitman criticized "the cell republic" in 1893 in a paper entitled "The Inadequacy of the Cell-Theory of Development." Just two years earlier, in his Woods Hole lecture, he had argued that organismic and human social structures were similar. But Whitman now denied that association and protested against the view that organismic organization could be explained in terms of reciprocal interaction of parts and against extending the struggle for existence and mutual interdependence to the development of higher organisms. Dismissing such accounts as based on "anthropomorphic conceptions," he now sought principles of structural unity.

Whitman's central argument was that cell division did not lead to organization; organization led to cell division. The characteristic growth of a complex organism was exhibited in the growing egg before it split up into cells. There was a preorganization in the egg cytoplasm itself which led to cell division. Cell division was an accompaniment, not a cause of differentiation. Therefore, cells were not the primary unit of organic structure. Cells did not make organisms; organisms made cells. That "the organism dominates cell formation," Whitman argued, was indicated by comparative embryology: one could find organisms using for the same purpose one, several, or many cells, massing its material and directing its movements, and shaping its organs, "as if cells did not exist, or as if they existed only in complete subordination to its will, if I may so speak."[60] In protists, complex organizations were worked out within the limits of a single cell. Cell theory did not apply at all to "unicellular organisms," in which it was obvious that cell division was the result, not the cause of organization or structural duplication.

The following year the Cambridge biologist Adam Sedgwick published a paper entitled "On the Inadequacy of the Cellular Theory of Development." In it, he attacked cell theory, claiming that it "blinds men's eyes to the true relations of cell organization and ontogeny."[61] Sedgwick's convictions found several followers in the twentieth century.[62] One of the most outspoken critics of cell theory was a former student of his, Clifford Dobell (1886–1949).

Dobell took the criticisms of cell theory one step further to insist that one simply could not use protists to study the origins of the metazoan cell. He had visited Hertwig's laboratory in 1907 in Munich and four years later launched an attack on German cell theory in a paper entitled "The Principles of Protistology." "The present paper," he warned, "is largely analytic and destructive: but it is so of necessity for it is useless to attempt to build on a rotten foundation."[63] The "rotten foundation" was the view that multicellular organisms were cell communities, and that the Protista were primitive unicellular organisms. He asserted that this anthropocentric view had only served to freeze studies of the Protista in their tracks: "The great importance of the Protista . . . lies in the fact that they are a group of living beings which are organized upon quite a *different principle* from that of other organisms."[64]

Protists were acellular organisms, and cells were not organisms. There were no *unicellular* "protozoan" precursors of metazoa. Dobell insisted that protists were just as far removed phylogenetically and temporally from the hypothetical unicellular progenators of metazoa as were the so-called higher organisms. The "protozoa to man hypothesis" rested on the biogenetic law that ontogeny was the recapitulation of phylogeny. As applied to cell theory, recapitulation theory assumed that when the egg undergoes segmentation in ontogeny, it repeats the process which occurred in phylogeny when the Metazoa arose from "unicellular" ancestors.[65]

But Dobell denied any "real analogy" between an egg dividing into two blastomeres and a protist dividing into two protists. An egg dividing was not one whole organism becoming a colony of several whole organisms, each of the same value as the original egg. A protozoon undergoing division was one whole organism becoming two whole organisms of the same value as the original whole organism. He reasoned, if segmentation of a metazoan egg were really analogous to the divisions of a protozoon, it would produce a cluster of eggs and not a differentiated organism.[66]

Minchin could not have disagreed more. He recognized that calling protists "unicells" was anthropocentric. So long as the Protozoa were studied entirely by themselves, without reference to any other forms of life, he agreed, they may be termed noncellular in the sense that they are not composed of cells. However, when they were compared with multicellular organisms, the term "unicellular" was perfectly applicable. He upheld the concept of the cell state, asserting that the "cells" composing the complex body of the higher animals and plants must be regarded as living individual organisms that are mutually interdependent. There was a homology between the Protozoan and the body cell of the Metazoa. The only difference he could see between them was that the former may reach no higher degree of complexity than the single cell.[67]

The arguments Minchin put forward for the organismic status of cells were of the same kind for demonstrating intimate symbiosis: tissue cells removed from the body could live and multiply in artificial culture media. Moreover, some of the cells of all metazoa permanently retain their complete independence of movement and action. This was especially apparent in the simplest and most primitive types of metazoa such as sponges and coelenterates, which Minchin studied. In the course of ontogeny entire groups of cells could alter their relative positions in the body as the result of migrations performed by individual cells. It was also well-known that if the adult sponge or hydroid was broken up completely into its constituent cells, those cells can come together again and build up, by their own individual activity, the regenerated body of the organism. For these reasons, it seemed to Minchin to be impossible to regard the body cells of the Metazoa otherwise than as individual organisms complete in themselves, primitively as independent as the individual protozoon, and in every way comparable to it.[68] There were also, of course, striking structural similarities of protozoan and metazoan cells. The protozoan cell was differentiated into nucleus and cytoplasm; in many cases the structure of the nucleus was virtually identical to that of the nucleus of metazoan cells. Minchin was at a loss to conceive what further criteria of homology between a protozoan and a metazoan cell could be demanded.[69]

But even if one agreed that protozoan and metazoan cells were homologous, there was still one other objection to studies of cell origins: one could never really be certain of what events had actually taken place in the remote past. In Minchin's words, they could only be "inferred dimly and vaguely from such fragments of wreckage as are found stranded on the sands of time in which we live."[70] Nonetheless, he insisted that the evolution of the cell could be investigated as a morphological problem of the same order as that of the phylogeny of any other class or phylum of living beings—and by the same methods of inquiry. First, there was the comparative method, whereby different types of cell structures could be compared to one another in order to determine what parts were invariable and essential and what were sporadic in occurrence and of secondary importance, and if possible to arrange the various structural types in one or more evolutionary series. Then there was the ontogenetic method: the study of the mode and sequence of the formation of the parts of the cell as they come into existence during the life history of the organism. One could also investigate the chemical evolution of the organisms. This method had already been applied to bacteria, which were characterized and classified largely by their chemical activities; it had yet to be applied to the more complex protozoa.[71]

Minchin recognized that these methods offered no guarantee that the phylogenies and evolutionary series one constructed were absolutely true. But he insisted that these investigations were useful in other ways than "as attempts to discover truth" or "striving towards a verity which is indefinable." Even if his own scheme for the evolution of the nucleated cell were but "a midsummer-night's fantasy," he claimed that it coordinated a number of isolated and scattered phenomena into an orderly and intelligible sequence, which at least allowed a comprehensive view of them. Rival theories would be more or less useful in accordance with their ability to correlate more, or fewer, accumulated data. As Minchin saw it, even if his address succeeded only in arousing interest and reflection, and in stimulating inquiry and controversy, it would have fulfilled its purpose.[72]

A Midsummer-Night's Fantasy?

Minchin's evolutionary search started with the primary structural differentiation of the typical cell: the distinction of the nucleus and cytoplasm. He began with the suggestion that the cytoplasm and nucleus represented two distinct organisms which had entered into symbiosis in the remote past, a view he attributed to Boveri and Merezhkovskii.[73] He did not challenge the view that the chromosomes were comprised of minute primitive organisms. In fact, he argued at length that "the extraordinary powers and activities exhibited by the chromatin" could be explained only on the hypothesis that "the ultimate chromatinic units [were] independent living beings, as much so as the cells composing the bodies of multicellular organisms."[74] He did, however, object to the notion that "true cells"—cells with nucleus and cytoplasm—emerged from a symbiosis between two corresponding kinds of simple organisms, mycoplasm and amoeboplasm, as Merezhkovskii had called them.

Merezkhovskii's speculations were open to criticism from several points of view. Minchin doubted the validity of his classification of bacteria, cyanophyceae, and fungi as a group distinct from all other living beings. He pointed out that botanists were in general agreement that fungi possessed nuclei similar to those of higher plants. He also denied any evidence for the existence of any organism lacking chromatin.[75] Indeed, to Minchin the weakest aspect of Merezhkovskii's scheme was the lack of any evidence that organisms consisting of cytoplasm (amoeboplasm) alone could ever have existed. He dismissed the notion of a primitive undifferentiated protoplasmic substance as proposed by Haeckel, pointing out that all those organisms referred to by Haeckel as such had been found to consist of ordinary cytoplasm containing nuclei or nuclear substance (chromatin).[76]

Minchin attributed immortality to the nuclear chromatin particles and declared that they were "the only constituents of the cell which maintain persistently and uninterruptedly their existence throughout the whole life cycle of living organisms universally."[77] In making this "apparently sweeping and breathless generalization," he was aware of claims for other self-perpetuating constituents of some cells. There were statements that centrosomes (or centrioles) were self-perpetuating cell constituents; but whether that was true or not, it seemed quite certain that there were organisms that did not possess centrosomes at all. There were also assertions that mitochondria were persistent self-reproducing elements and that they were bearers of heredity. But again, Minchin could cite conflicting reports, and he claimed that there was no evidence that they were of universal occurrence in the Prostista. There was almost indisputable evidence that chromatophores were self-reproducing, but they did not occur in all organisms. Only the nucleus and its contained chromatin elements, the chromosomes, were universal. Indeed, Minchin was prepared "to challenge anyone to name or discover any cell constituent, other than the chromatinic particles, which are present throughout the life cycle, not merely in some particular organism, but of organisms universally."[78]

All of Minchin's arguments led to the conclusion that the earliest living beings were ultramicroscopic particles of the nature of chromatin. He employed Merezhkovskii's term "biococci" for these hypothetical primitive organisms.[79] They had to be free-living organisms capable of building up their bodies by synthesis of simple chemical compounds. Thus he offered his own reconstruction of the sequence of events leading to the emergence of the metazoan cell without recourse to symbiosis. From the primitive biococci type two new types of organisms arose; one continued to specialize in the vegetative mode of life characteristic of the biococci, while the other developed an entirely new habit of life, a predatory existence. In the vegetative type, the first step was the development of a rigid envelope around the body resulting in the bacterial type of organism. From this familiar bacterial type an infinity of forms arose by processes of divergent evolution and adaptation.

In the evolution from the biococcus to the predatory type of organism, the chief event was the formation of a surrounding matrix of protoplasm. This could have occurred around a single individual biococci or by the formation of zoogloea-like colonies. It resulted in "the brand of Cain, the prototype of the animal," a class organism which could sustain itself only by capturing, devouring, and digesting other living organisms. The streaming movements of the protoplasm enabled it to flow round and engulf other creatures; the vacuole formation in the protoplasm enabled it to digest and absorb the substance of its prey by the help of ferments secreted by the chromatin grains (biococci).[80] The next stage in evolution was the organization of the chromatin grains into a definite cell nucleus. Minchin believed that one could observe this process actually taking place in many protozoa in which a "secondary" nuclei arises from extranuclear granules of chromatin, which Richard Hertwig had named *chromidia* in 1902.[81]

During the first three decades of the century, chromidia captured the interest of many leading cell biologists and protistologists, and it was discussed in virtually every leading cell and protozoology text of that time.[82] Chromidia could be observed in several groups of protozoa; it could be found scattered in the cytoplasm, or aggregated in certain regions of the body to form "chromidial masses" or "chromidial nets." In some cases it appeared as extrusions from the nucleus. In others the nucleus completely fragments and ceases to exist as a definite nucleus, resolving entirely into chromidia. In still other cases chromidia arose from preexisting chromidia, by growth and multiplication, thereby maintaining a chromidial mass or stock which was propagated from cell to cell through many generations in addition to the nucleus. In some species, a true nucleus was only temporarily absent during certain phases of the life cycle; all the chromatin was then found in the form of chromidia from which nuclei arise. In Minchin's view this represented a temporary reversion to a more archaic and ancestral condition.[83]

The formation of the nucleus was the starting point of another series of complications and elaborations in cell evolution. This stage reached its highest development with the evolution of nuclear division by karyokinesis (mitosis), which ensured exact quantitative and qualitative distribution of chromatin among cell generations. Minchin assigned enormous importance to the evolution of mitosis: the "very existence of multicellular organisms composed of definite tissues, is impossible until the process of karyokinesis had been established and perfected." In order to have tissue formation it was essential that all the cells which build up any given tissue be similar, practically identical in their qualities. Therefore, he reasoned that the exact qualitative division of the chromasomes as effected in mitosis was an essential condition.[84]

Not being a botanist, Minchin approached the question of the divergence of animal and vegetable cells with caution. But as a protozoologist, it was clear to him that the typical green plant cell originated among "flagellated protozoa." Some protozoa had acquired chromatophores which enabled them to abandon the animal mode of life in exchange for a vegetative mode of nutrition by means of chlorophyll corpuscles. However, Minchin did not feel competent to discuss the question of whether or not chromatophores arose as symbionts.[85]

Missing Mitochondria

Minchin's address did not have much success in spurring biologists' interest in the origin of cells and the natural history of the Protista.[86] The study of protists and bacteria remained largely attached to problems of disease. However, Minchin's challenge to name any other cell constituents besides the chromosomes that were permanent organs in all cells was, in effect, met—with studies of mitochondria. These cytoplasmic constituents attracted the attention of botanists, zoologists, anatomists, physiologists, and clinicians, who all studied them from diverse points of view. A conservative estimate would place the bibliography of the second decade of the century relating to mitochondria at about five hundred papers scattered widely in the journals of many countries, especially France and Germany.

There were statements that mitochondria existed in the cells of all forms of plants and animals, and they had been repeatedly reported in unicellular organisms such as algae and fungi. Zoologists identified mitochondria in both eggs and sperm and some leading cytologists ascribed to them the power of independent growth and division and considered them to represent a mechanism of heredity comparable to chromosomes. They were often considered to be fundamentally important both in the chemical activities of the cell and for tissue development and differentiation, forming the source of many other cell components including chloroplasts. Mitochondria were observed and investigated under a variety of aliases: bioblasts, chondriosomes, plastidules, plastosomes, eclectosomes, and vacuolides.[87] The "correct" name for these granules was often the subject of heated discussions.[88] Richard Altmann was usually credited for discovering them, but this also became the subject of dispute, especially in France during the second decade of the century.

Those who studied mitochondria during the first decades of the century commented on why they had not been recognized and systematically investigated earlier with the classical cytological studies of the 1880s and 1890s. Some suggested that Altmann's assertions that they were "elementary organisms" existing in the form of colonies in cells had had a negative effect on their investigation and deterred many from studying them.[89] Cytologists had also been dazzled by the dance of the chromosomes. The subsequent rise of genetics and the Mendelian chromosome theory, during the first decades of the century, further blinded biologists from recognizing other important cellular constituents. But still there was more to it than the theory-ladenness of observations and the tendency of scientists to "see what they want to see." Observations are also technique-laden.[90] This is especially important for understanding the plight of mitochondria.

In most cases, mitochondria could not be investigated or even seen by the ordinary cytological techniques. Worse still, the techniques for fixing and staining chromosomes actually precluded observations of mitochondria. With the attention concentrated on the nucleus, the aim in making up fixatives was to show nuclear detail. For this purpose mixtures containing alcohol, chloroform, or acetic acid were employed because of their rapid penetration. But these substances destroyed mitochondria. The more attention was focused on the nucleus because of its assumed central role in heredity and development, the less chance there was for observing mitochondria. A vicious cycle was maintained.[91]

This is not to suggest that the renewed interest in and development of research on mitochondria was simply technique-driven. The possibility that mitochondria played an important role in heredity and development was a great stimulus to their study. In 1908, the acclaimed German cytologist Friedrich Meves at Kiel stated that all cellular differentiations were formed from mitochondria and that mitochondria constituted part of the material basis of heredity.[92] Many European cytologists supported the "eclectosome theory" of the French histologist Claude Regaud at the Pasteur Institute, according to which mitochondria select substances from the surrounding cytoplasm and condense and transform them into diverse products.[93] These kinds of statements attracted worldwide attention, coming at a time when there were serious protests against the role that had been often assigned to the chromosomes in heredity and development.

With the rise of genetics, much attention was focused on the transmission of Mendelian genes from one generation to the next. But the first generation of geneticists virtually ignored the problem of cellular differentiation and morphogenesis. Although they said nothing about how genes could direct ontogeny, leading American geneticists insisted that one day they would demonstrate how the chromosomes governed development. Moreover, they persistently maintained that there was virtually no evidence for cytoplasmic inheritance in their crosses.[94]

Many of those who opposed the exclusive or predominant role of nuclear genes in heredity and development argued that Mendelian geneticists should not be expected to observe cytoplasmic inheritance by the techniques they used. Geneticists were concerned not with similarites but with differences between individuals of a species. Mendelian genetics therefore seemed to deal with rather trivial traits such as eye color and hair color, not with the fundamental characters that distinguished phylogenetic groups. That the cytoplasm was responsible for fundamental characteristics (macroevolution) whereas the nucleus was responsible for traits which did not extend beyond the species (microevolution) was widely maintained by embryologists in many countries as well as by neo-Lamarckian biologists, especially in France.

The possibility that development was regulated primarily through the cytoplasm and not the nucleus was a great impetus to the study of mitochondria. As the Canadian histologist E. V. Cowdry put it in 1918, "There is nothing new in the conception that there is such a thing as cytoplasmic heredity. . . . What is new is the view that mitochondria carry it."[95] But the evidence for the role of mitochondria in development and heredity did not come solely from studies of zoologists. Some of the strongest support for the view that mitochondria were formative granules came from their study in plants led by Alexandre Guilliermond in Lyon, who championed the view that plastids in plants actually originated from mitochondria. At the same time, many of those who studied mitochondria repeatedly commented on their similarity to bacteria.

Merezhkovskii denied that plastids originated from mitochondria, and he dismissed as unreasonable the idea that mitochondria arose as symbionts. He viewed such ideas as wholly antagonistic to his theory of "the symbiotic nature of plants" but also his "theory of the two plasmas." As he saw it, "all of his theory of symbiogenesis would fall in ruin."[96] Despite Merezhkovskii's denial, the possibility that mitochondria were indeed nothing other than symbiotic bacteria was developed into a comprehensive theory by the French biologist Paul Portier, whose experiments and theorizing on *Les Symbiotes* caused great controversy in France in the years following World War I. And it is to this work that we shall now turn.

6

Les Symbiotes and Germ Theory

All Living beings, all animals from Amoeba to Man, all plants from Cryptogams to Dicotyledons are constituted by an association, the "*emboîtement*" of two different beings.

Each living cell contains in its protoplasm formations which histologists designate by the name of "mitochondria." These organelles are, for me, nothing other than symbiotic bacteria, which I call "symbiotes."

PAUL PORTIER, 1918[1]

Pasteur effectively quieted all proponents of spontaneous generation.[2] Microorganisms in the tissue of animals were not the consequence, but rather the cause of disease. This work led to the isolation and systematic study of germs, to vaccination, and to the preparation of preventive and curative serums. The germ theory of disease transformed medicine and part of biology. At the same time, however, it led to the dogma of asepsis of normal healthy animal tissue. Organic liquids such as blood, urine, and lymph collected with necessary precautions against introducing extraneous microorganisms do not give rise to germs. Therefore, the tissues of healthy animals were deprived of germs. If microorganisms appeared, it was due to a technical fault: perhaps the operating instruments, the flask in which they had been deposited, or the air which penetrated into the flask had carried these germs. Otherwise, it was because the animal was stricken by a *parasitic infection.*

Innumerable experiments made each day in bacteriological laboratories, at the Pasteur Institute and elsewhere, confirmed the dogma of the asepsis of healthy tissue. There were, of course, some well-known exceptions: nitrogen-fixing "bacteroids" in the root nodules of legumes, symbiotic fungi of orchid bulbs and forest trees, unicellular algae living in "lower animals" from protozoa to sea anemones. These and other cases were investigated and/or discussed by prominent biologists in France, including Félix Henneguy, Raphaël Dubois, Yves Delage, Marie Goldsmith, Félix Le Dantec, Gaston Bonnier, Noël Bernard, Maurice Caullery, and Félix d'Herelle during the first three decades of the century. Such cases were well-known, but they were tolerated by bacteriologists more than they were accepted, and then again they occurred in the "lower orders." Yet animals live in a universally contaminated world. All objects that surround us are covered with microorganisms. The air we breath is charged with them; the foods we ingest are populated with them. Was it really possible to have life without them? Could animals be constituted without them?

This was the background against which Paul Portier (1866–1962) published his book *Les Symbiotes* (1918) and launched his far-reaching theory which held that, apart from bacteria, all organisms were double beings formed by the association of two different kinds of creatures. Every cell contained partner bacteria, indispensable for effecting organic synthesis, which Portier called *symbiotes*. Portier argued that bacteria alone were autotrophic, that is, capable of feeding independently. All cells, and consequently all animals and plants with a cellular constitution, would be heterotrophic and would assimilate only through the intermediary of symbiotic bacteria. He supported his belief in the universal presence of bacterial symbionts with the claim that aseptic life was virtually impossible. It was well-known that the intestines of animals usually contained a rich and varied "flora and fauna," sometimes useful and sometimes harmful, and that microorganisms in the paunch of ruminants played an important part in the transformation of food substances.

But Portier went much further and asserted that he was able to cultivate bacterial symbionts from organs of all types in a wide variety of vertebrates. He described the individual strains and made attempts to find the microbes in the cells. But if all this was so, then his theory faced the apparent paradox that symbiotic bacteria had not been seen before in all kinds of plants and animals. His response to this objection was that, in fact, such symbiotic bacteria had often been seen, but observers had simply not recognized them as such. They were "mitochondria," those minute formed bodies, with many an alias, which had been described in the cytoplasm of all sorts of cells. Based on his identification of mitochondria as bacteria, Portier constructed an all-encompassing theory of symbiotes, elaborating their possible role in metabolism, nutrition, heredity, development, cancer, parthenogenesis, and the origin of species. *Les Symbiotes* rose to the center of heated discussion in France—hailed by some as marking a scientific revolution, criticized by many others, ridiculed, to be forgotten and finally dismissed from the annals of science.[3]

Neo-Lamarckism, Infection, and Immunity

Studies of the relations of higher organisms to microbes in France were not only carried out with respect to bacteriological doctrines of asepsis and cytological investigations of mitochondria, chloropasts, and other cytoplasmic bodies; they were also studied as part of the general interest of French biologists in the effects of the environment on hereditary variation. The origin of variations due to microbes in the environment was of considerable interest when the Fourth International Conference on Genetics was held in Paris in 1911.[4] The leading French presentation was made by André Delcourt and Emile Guyénot. They argued that by not keeping the environment controlled in their *Drosophila* experiments, Mendelian geneticists in the United States had placed themselves in a weak position. From an evolutionary point of view, nothing could be concluded from their work, especially as it concerned the origin of variations.

Delcourt and Guyénot had developed techniques by which temperature and moisture were controlled, the environment was kept sterile, and the food given to the flies was analyzed and kept constant. As Limoges has remarked, "Their research program was clearly Lamarckian, with the expectation that studying the effects of the fluctuation of single variables of the environment would enable them to attribute the source of hereditary variations to environmental influences."[5] Though they had high expectations, by 1917 none of the anticipated results had been obtained. Guyénot became a convinced Mendelian and pleaded for a mutationist, non-Lamarckian viewpoint for the nature of inherited variations in multicellular organisms.[6] In 1924 he published the first textbook on classical genetics in French.[7]

Delcourt and Guyénot were not alone in considering variations due to the effects of microbes. Noël Bernard attributed large, sudden inherited variations in plants to parasitism and subsequent symbiotic relationship of parasite and host. Bernard became well-known for his investigations of mycorrhiza and orchids during the first decade of the century. In fact, his studies of this relationship, published in 1902, became the exemplar for understanding the origin and underlying dynamics of symbiosis involving microorganism and the characteristic changes in the host which result from it.

Bernard associated his work with bacteria on the root nodules of legumes and with fungi in the roots of forest trees: the "common life of animals and higher plants" with microbes (algae, bacteria, or fungi) "appeared more and more to be a phenomenon of a great generality."[8] It was rare for an organism not to be infected at some time in its life by some microorganism. Most often this was *accidental;* that is, one individual of a species was infected, giving rise to disease. But the foregoing examples represent cases where infection was *normal*—all individuals of a species were infected—and the characters which appeared in consequence to the infection had to be considered as *specific characters.* In Bernard's view, they were often considered hereditary characters and therefore their importance as symptoms of normal infection often escaped biologists' attention.[9]

Yet Bernard was reluctant to employ the term symbiosis since it "had deviated from its etymological meaning," to "imply a useless finalist hypothesis."[10] The useless finalist hypothesis was "symbiosis . . . taken in the sense of the association of two specifically distinct beings which harmonize their functions for the greater good of the community."[11] In his view, the distinction between the terms "parasitism" and "symbiosis" only reflected the extent to which studies of symbiosis had been divorced from studies of the pathological effects of microorganisms in "higher" animals. "One has a tendency", he wrote in the *Bulletin de l'Institut Pasteur* in 1909,

> to separate cases of "symbiosis" from cases of "disease" and to study them from completely different points of view. I will try to the contrary, starting

with symbiosis, to understand disease, and it is appropriate first of all to show, by examining a classical example, that there is no absolute distinction to be made between these two orders of phenomena.[12]

Botanical studies of symbiosis, beginning with the lichen, paralleled, but had virtually never interacted with research on the development of germ theory of disease. During the 1860s Pasteur had devised methods for controlling contagious diseases and attempted to isolate the microbes of the silkworm disease which was seriously affecting French industry. Robert Koch worked on the etiology of anthrax and in 1876 was able to demonstrate the existence of a specific pathological organism, the anthrax bacillus.[13] This widely acclaimed finding led to the rapid identification of other bacterial pathogens during the 1870s and early 1880s: *Staphylococcus, Neisseria gonorrhoeae, Salmonella typhi, Streptococcus,* and *Mycobacterium tuberculosis.*[14] These studies closely preceded Elie Metchnikoff's phagocytosis theory, according to which natural immunity was reduced to the direct action of phagocytes, which engulf and digest invaders.[15]

Metchnikoff's theory of immunity had never been extended to plants. As Bernard saw it, the extent to which studies of symbiosis between microbes and higher plants had been separated from studies of disease and immunity was well exemplified in A. B. Frank's suggestion that forest trees had developed a mechanism for attracting and capturing the beneficial fungi to procure nitrogen. "This curious hypothesis" Bernard remarked, "is symptomatic of the state of mind with which one has most often approached the study of symbiosis."[16]

Bernard showed that the fungi played an important role in initiating germination of orchid seeds. But the premise for his work was that the fungus and the orchid were in conflict, and he aimed to reveal "the means of attack and defense of the two antagonists."[17] In his view, the "first impression of harmony" was due to superficial examination of the end result of the association; it disappeared, "at least in part, when one examines the way in which the association comes into being."[18]

By inoculating seeds of the same orchid with fungus from different species, Bernard showed that the fungus penetrates the seed and one of three results may ensue: the resistance of the seed may overcome the virulence of the parasite and the latter is destroyed; or the virulence of the fungus outweighs the resistance of the seed and the seed is destroyed; or, finally, the virulence of the one and the resistance of the other balance, and a symbiosis results, which is, in effect, a constrained parasitism.[19] "An impartial examination of these facts," he wrote, "reveals clearly that symbiosis is an exceptional state, rarely accomplished and bound together by gradual transitions of an infectious disease under its diverse forms."[20] Thus when one studies this association from "an experimental point of view," he declared, one sees "not only that symbiosis is *à la frontièr de la maladie,* but also that its study is the *terrain de choix* for understanding the laws of plant pathology."[21]

Portier's work on symbiosis was also preceded by his work on immunity, or rather increased sensitivity, or anaphylaxis. However, unlike Bernard,

Portier did not believe that all cases of symbiosis originated from infectious disease or was an "exceptional state." Indeed, not only were his views on symbiosis situated precariously on the edge of mainstream French biology by 1918—this was also true of Portier himself.

The Prince and the Pauper

Portier had become a well-known figure in the French biological and medical communities during the first decade of the century. He continues to be known today chiefly for his work with Charles Richet (1850–1935) on the experimental induction of hypersensitivity, or anaphylaxis.[22] The discovery of anaphylaxis changed medical and physiological perspectives on the principles of immunization. The work which led to this discovery was carried out on board *Princesse Alice II,* a scientific yacht of Prince Albert I of Monaco, who is celebrated as one of the "patron saints" of oceanography.[23] During his reign from 1889 to 1922, the prince outfitted a series of ships and made numerous scientific expeditions. Not only did the prince personally engage in these oceanographic studies, he also founded two institutions for oceanographic research and teaching: the Musée Océanographique de Monaco opened in 1899, Jacques Cousteau was later a world-famous director, and an affiliate institution, the Institut Océanographique de Monaco, opened in Paris in 1906.[24] The prince also sponsored and encouraged much of Portier's early interest in physiology and symbiosis.

Portier came from a family that traditionally served in the national bureaucracy.[25] When he was in his early twenties he was offered a career in the Ministry of Finance. However, he had no taste for his work at the Bureau de la Direction des Contributions and maintained a deep interest in biology. His family allowed him to follow his own inclination, provided that he first study medicine. While at the ministry, he took courses at the Sorbonne in preparation for the *licence ès sciences,* which he completed in 1891 after attending the courses of eminent biologists, including Henri Lacaze-Duthiers, the founder of the laboratories at Roscoff and Banyuls, who taught Portier to observe marine animals in their environment. Portier completed his medical studies in 1897, but he never practiced medicine. While completing his medical degree, Portier maintained contact with research laboratories first at the Collège de France, where Félix Henneguy taught him histological techniques. Then, beginning in 1894 he worked as an assistant in the physiology laboratory of Albert Dastre, who held the chair of physiology at the Sorbonne. It was there that he first became acquainted with Prince Albert, who, in 1898, invited him to be a member of a small group of scientists on his scientific cruises on the *Princesse Alice II.*

Portier's celebrated studies with Charles Richet began in the summer of 1901 during a cruise in the Mediterranean Sea. The prince and his scientific director, Jules Richard, had suggested the topic for the research that ultimately led to their experimental induction of hypersensitivity. During a previous cruise, Prince Albert and Richard had become interested in the way

Physalia, a genus of the Portuguese man-of-war, captured its prey. Sailors were familiar with sharp pain from touching the tentacles. The prince and Richard suspected that a powerful poison secreted by the tentacles could account for the phenomenon and proposed to Portier and Richet that they study the poison.

In 1913, Richet was awarded a Nobel Prize for the discovery of anaphylaxis. Portier, though the principal author of the papers describing the discovery, received only a passing mention by Richet on the occasion of the acceptance. This apparent injustice has not gone without comment from scientists. It has been argued that Portier's failure to receive adequate recognition "is in keeping with a familiar academic pattern: an older, distinguished figure may be allowed to overshadow a younger, relatively unknown person."[26] At the time of their discovery, Richet was fifty-two years old and had been professor of physiology of the Medical Faculty at the University of Paris for fourteen years. He had already achieved prominence as a research scientist, had been elected to the Académie de Médicine, and was the author of numerous publications on subjects ranging from psychosomatic medicine, physiology of muscles and nerves, diuretic effects of sugars, calorimetry, to therapy of infections and cancer with antisera. In addition, he edited an eight-volume encyclopedia of physiology and wrote several scientific monographs.

Portier was thirty-six years old, had received his degree as doctor of medicine only four years previously, and was a humble laboratory assistant at the Sorbonne at the threshold of his academic career. It has been recently claimed that Portier was never bitter about the lack of recognition accorded him, and he and Richet remained good friends. Because of his "gratifying personal achievements, he was content to see Richet receive the bulk of recognition and the Nobel Prize for their joint discovery without envy or rancor."[27] However, these are virtues usually accorded saints, and no matter how gracious and humble Portier may have been, his experience with Richet made him ensure that his originality and intellectual property rights be protected for his work on *Les Symbiotes.*

Portier's collaboration with Richet ended with the publication of their discovery of the experimental induction of hypersensitivity in 1902, but his association with the prince continued. When the Institut Océanographique de Monaco in Paris was opened, the prince created a chair of physiology and appointed Portier as professor. This position brought him in contact with many leading European biologists. Between 1917 and 1920 some two hundred researchers visited Portier's laboratory of physiology, working on diverse problems ranging from circulation of sea plankton, to sense of balance in fishes, to parthenogenesis in sea urchins, to symbiosis of marine algae and sponges.[28] Portier's own work was equally diverse. His studies of marine life, from bacteria to vertebrates, oriented him around comparative physiology, a discipline which at the time had hardly been cultivated in France. During the first decade of the century, he published papers on the glycolysis of sugars and endocellular enzymes of mammal organs; he also developed techniques for collecting marine bacteria and studied their distribution and physiology. In short, the position Prince Albert created for him allowed him to combine natural history with laboratory experimentation. Portier dedicated *Les Symbiotes* to Prince Albert I of Monaco and to comparative physiology.[29]

Challenging the Doctrine of Asepsis

When he had been appointed professor of physiology at the Institut Océanographique, Portier had held only a doctorate in medicine. The work for his *doctorat ès science naturelles, Physiological Research on Aquatic Insects,*[30] begun in 1905 and completed in 1911, was the starting point for his studies of symbiosis. In the course of his studies, he was struck by the frequent occurrence of microorganisms in the larvae of xylophagous insects—insects that eat inert substances such as cellulose and wood. The infection of these insects was well-known but had been considered to be a phenomenon of parasitism. Portier, however, took a different perspective: the microorganisms were symbionts. He studied the development of the caterpillar *Nonagria typhae,* which he believed enclosed the spores of the fungus *Isaria.* He found the fungus in all the tissues of the adult, including the egg. He used this evidence to address one of the central problems of his day: Are microbes necessary for the life of higher animals and for humans?

The digestive tube of virtually all animals was populated so soon after birth that many bacteriologists wondered if these microorganisms did not have an important, or even indispensable role in the transformation of food substances, as they did in herbivores and in ruminants. Our food, having been absorbed, has to be profoundly transformed from a chemical point of view. Glands next to different segments of the digestive tube (salivary, gastric, intestinal, pancreatic) were known to secrete liquids rich in diastases and other enzymes. But microorganisms were also known to be great producers of these chemical agents. What was their role in the transformation of food? How could one even decide if the gland was not borrowing these enzymes from the microorganism? There seemed to be only one way to find out: take an animal at birth, place it in a sterile enclosure, and nourish it with food that was free of germs. This would be a very difficult and delicate experiment to perform.

Since the early 1880s, Pasteur himself had been preoccupied with this question. Though he believed that the blood, tissues, and embryos in utero and in eggs were free from bacteria, he postulated that the existence of microbes in the intestines was absolutely necessary for the host. He first publicly expressed the idea of rearing germ-free animals after presenting work on nitrogen fixation with sterile and innoculated plants by his student Emile Duclaux to the French Academy of Science in 1885: "Without wanting to affirm anything, I do not conceal the fact that if I had the time I would undertake this study with the preconceived idea that, in these conditions, life would be impossible."[31]

Pasteur, nearing the end of his scientific career, never undertook this study. In the years that followed, the question of whether it was possible to make germ-free animals remained the subject of debate in Europe. There were several attempts to carry out the definitive experiments, but the results were confusing. Guinea pigs raised aseptically seemed to gain weight, chicks lost weight, tadpoles survived but not very well.[32]

In his studies of insects, Portier found another way of approaching the problem. There were tiny larvae of certain butterflies which lived in aseptic conditions inside certain leaves, devouring parenchymal cells of the leaf. An examination of their tissues and excrement showed in absence of bacteria. Other caterpillars lived the first phase of their life in aseptic conditions but later lived in a contaminated environment. An examination of their digestive tube revealed an abundance of microorganisms. It seemed to Portier that they were necessary for digesting the cellulose wall of the leaf and releasing the contents of the cell for the use of the larvae. This seemed probable in light of the important role played by microorganisms in the digestive tube of herbivores and in the paunch of ruminants.

In further studies of xylophagous insects, Porter reported that while most of the microorganisms were devoured by the larvae, some spores could be found in the blood and became lodged in the tissues of the larvae, where they could be detected by appropriate histological techniques. Some could migrate to the eggs, and they multiply as the egg divides so that the young larvae are already provided with their own symbiotic microorganisms, without which they could not survive. This was a case of "hereditary symbiosis." In demonstrating the presence of microorganisms in healthy tissue, Portier recognized all too well that he was moving upstream against the doctrine of the asepsis of healthy tissues. Animals were often metaphorically regarded as "fortresses" with many walls to protect them from the invasion of vile microbes. If the first barriers were broken through, if the enemy forced its way in, it is in most cases captured and digested by mobile cells, the leukocytes.[33] There was no normal passage of microorganisms into the tissues of organisms.

Portier asserted that certain microorganisms in xylophagous insects typically escaped this destruction. Among all the tissues, those which showed the most remarkable invasion of microorganisms were the fatty tissue annexed to the genital organs, particularly those around the testicle. In other insects, the phenomenon was less marked, though he could find mitochondria enclosed in the cytoplasm of their cells. Portier had read histologists' remarks about the resemblance of these cytoplasmic inclusions to bacteria or yeast. Similar inclusions called "bactéroïdes" were noticable in certain cockroaches. In any case, the presence of microbes in the fat tissue next to the genital organs of insects incited Portier to try to determine whether the genital organs of other animals, even vertebrates, normally contain microorganisms, and if they did, whether they might also play important roles in the metabolism of the organism.

When World War I broke out, Portier, at age forty-eight, was released from all military obligations. Nonetheless, he asked to serve and was assigned to a temporary hospital at Bar-sur-Aube. He held a medical degree, but he was not a medical practitioner. In order to be useful, he established a bacteriological laboratory with financial support from Prince Albert of Monaco. In 1915, he published a paper on the resistance of certain varieties of bacteria to chemical agents and another on the presence of micrococci in the blood of typhoid victims coming from the front.[34]

Portier's first results on symbiosis in vertebrates were published in 1917. He began by listing studies in France which indicated the important role of symbiotic microorganisms in plants and animals. Beginning in the 1880s, Raphaël Dubois in Lyon had studied the phenomenon of luminescence in animals and had suggested that symbiotic microorganisms could intervene in this process. He located the seat of animal light- and color-producing organs in organelles he called *vacuolides,* which he later identified with mitochondria.[35]

Dubois's studies were taken up by the Italian biologist Umberto Pierantoni (1876–1959) at the Zoological Station in Naples.[36] He demonstrated that the cells of the light organs of certain beetles and cephalopods contained luminous microorganisms that also existed in the egg and could be transmitted from generation to generation. Designating this phenomena "hereditary symbiosis," he asserted that the "vacuolides" of Dubois were bacteria adapted to intracellular life. Pierantoni was well-known for his claims that luminescence and pigmentation were due to symbionts.[37] He also carried out experiments in order to prove by cultures that the pigment granules were symbiotic bacteria. In certain cases he was able to obtain cultures of these microorganisms and an emission of light in vitro. All this, he argued, "would assign to the cytoplasmic inclusions and perhaps to many constituents of the protoplasm an autonomous life and a specific activity, to the benefit of the organism in which they live."[38]

There were studies of nitrogen-fixing bacteria in the root nodules of legumes carried out at the Pasteur Institute by the bacteriologist Pierre Mazé,[39] as well as Noël Bernard's famous work on orchids.[40] But there was precious little work on symbiosis in organisms "higher up on the scale of beings." Gabriel Bertrand[41] had isolated bacteria from the testicles of mammals. Victor Galippe[42] also claimed that he had frequently found microrganisms in the organs of reproduction and he associated them with the hypothetical entities which Pierre Béchamp had called "microzymas" in 1875.[43] Yet this work gained little recognition for the occurrence and significance of symbiosis in "higher animals."[44]

In order to convince biologists that microorganisms are normally present in healthy tissues of all animals, Portier reported that for twelve years he had been trying to obtain multiple proofs: morphological, bacteriological, and physiological. His aim was to culture the microorganisms in question, examine their biochemical properties, and compare them to those which were known to take place in tissue cells. This would be no easy task at the best of times. Many times his research was abandoned because of technical difficulties. In 1914 he was making headway when he was interrupted by the events of the Great War. Many of his cultures were lost, but he managed to reconstruct some of them.[45]

Portier reported that he had obtained from fat tissue of the testicles of vertebrates cultures of granules which transformed themselves into bacteria.

The microorganisms were all anaerobic; their form was extremely variable, and they were very resistant to physical agents such as heat, light, and ultraviolet rays, and to chemical agents, antiseptics in particular. They carried out the same processes known to occur in the tissue of vertebrates.[46] All exhibited the same physiological reactions: polymerization of sugars, utilization of nitrates for making organic nitrogen, decarboxylation, oxidation, and so on. However, Portier went further in his attempts to transform parasites into symbiotes, the vile into the benevolent. He associated his symbiotes with essential life-giving substances—vitamins.

Symbiotes and Vitamins

Physiologists of the nineteenth century had believed that a minimum of nitrogen, together with carbohydrates or fats to meet energy requirements, was enough to ensure the healthy and normal metabolism of an animal. But this view changed during the first decades of the twentieth century. Physiologists came to recognize that food must contain other particular substances of a chemical composition which remained unknown. They were given the name of vitamins (life-amines so-called because they were thought to be amines). A diet containing very small quatities of vitamins was indispensible for adequate nutrition. When these hypothetical entities were lacking, animals lost weight, and they soon died with a series of disorders discussed under the rubric of malnutrition, vitamin deficiency, or avitaminosis. Beriberi, scurvy of sailors and prisoners, and infantile scurvy were all considered to be aspects of this particular illness. Vitamins existed in the teguments of grains: chickens and pigeons fed dehusked rice got symptoms eqivalent to beriberi; mice died when fed on a diet of bread washed in alcohol. Vitamins also existed in certain animal fats (butter, egg yolk, cod liver oil) and were destroyed when heated to a temperature above 120°C.

In the course of his research on bacteria isolated from the tissues of healthy animals, Portier noticed some striking similarities between what he had learned about symbiotes and what was shown about vitamins. Symbiotes also existed in the tegument of grain, and they seemed to be absent in the central part of the grain; there was also an abundance of them in animal fats such as that contained in milk; and the temperature at which they were destroyed was very close to the reported temperature at which vitamins were altered. Portier carried out experiments on vitamin deficiency in pigeons to test his hunch that the microbes he isolated carried essential vitamins.[47]

Portier consolidated this work with his friend Henri Bierry, who was *Maître de Conference at École des Hautes Etudes.*[48] They injected massive doses of microorganisms extracted from the fat tissue of vertebrates into the venous system and cell tissue of vertebrates. The animals (dogs, rats, pigeons, frogs) survived the injections and remained in good health.[49] They were able to restore the health of malnourished white rats and pigeons fed on a diet deprived of vitamins by injecting large doses of appropriate symbiotic bacteria (microbes taken from the tissue of the animal in question and cultured on chemically defined media) under their skin.[50] The hypothesis that mitochondria were symbionts was saved for Portier's book; it appeared late in October 1918.

The Rhetoric of *Les Symbiotes*

In *Les Symbiotes,* Portier constructed a grand, all-embracing theory of symbiosis as a primordial characteristic of cell life. Bacterial symbiosis would be responsible for fundamental life-giving processes in all organisms. Bacteria would therefore be the only single organisms; all others would be doubles. "Avitaminosis" would be "asymbiosis." In the form of mitochondria, bacteria played an important role in fertilization, development, and heredity. Bacterial symbiosis was a source of evolutionary novelty and responsible for large sudden mutations; they were also important for understanding the causes of cancer. If right, he stood to change basic tenets of biology. If wrong. . . . In December 1917, Portier wrote to his friend and colleague Jules Richard, director of the Musée Océanographique de Monaco:

> I am putting the last hand to my book "Les Symbiotes." It will be completed in a few days. I put forth there all my ideas on the question. I synthesize not only my personal work, but also of many others which gravitate around the same problem, without their authors suspecting it, you understand.
>
> I deduce from all this a bold theory, and even one of excessive audacity. However, this is what I proclaim in the preface. But I produce such a great number of proofs of diverse nature that it seems to me absolutely impossible that my thesis is not true on the whole; naturally alternative experiments will modify some details, as for all theories.
>
> If I am right, it is a great wave of thought that will move across *all* of biology and which will instigate very numerous works.
>
> If I am wrong, I will bow and make my *mea culpa.* But not really, that's not possible.
>
> You know me. If I say this to you and in this way, I have to have very strong reasons; however, you can be the judge of that.[51]

Convinced that, once again, he had discovered something important, he wanted his intellectual property rights protected. If the bulk of the credit for the discovery of anaphylaxis had gone to Richet, Portier was not going to lose credit for the originality of his claims in *Les Symbiotes.* He made this clear when he sent his manuscript to Richard: "I want to disclose my work only to *you,* you will know why when you have read it. If I am not mistaken, there is something *very* important there, and this time I want to make sure that it belongs to me without possible contestation."[52]

Les Symbiotes was a masterful piece of rhetoric. Portier constantly alerted readers to his new perspectives on bacteria and comparative physiology, pointing out possible objections to his experimental results and his theory. He stated that his theory was "a veritable scientific heresy, and I admit that if these lines had fallen under my eyes a few years ago, I would have probably refused to take any further notice of the work which proclaimed it."[53] He persistently portrayed himself as his most serious critic, arguing that his conclusions were not driven by preconceived ideas but rather by the accumulation of facts over many long years: "I have begun with research on natural and permanent asepsis, and I have arrived without wanting to and almost despite myself, to the facts that I will present and the theory which relates to it."[54]

Portier argued against physiologists who agreed with the positivist philosopher Auguste Comte that "the animal life of man helps us to understand that of the sponge, but the opposite is not true."[55] He asserted that one should study the most different kinds of beings, and that such studies should revolve around comparative physiology. But it was the bacteriologists he relentlessly pursued, drawing them out with his rhetoric.

From beginning to end, he juxtaposed his views of bacteria to their medical perspective of bacteria as disease and their doctrine of asepsis of healthy tissues.[56] He repeatedly emphasized that his ability to cultivate microorganisms from the healthy tissues of animals was in direct contradiction to bacteriological doctrines. He asserted that when such cultures were produced by bacteriologists they were deemed faulty and dismissed as resulting from poor technique, from contamination. But Portier dismissed this a priori reasoning as being designed only to support the doctrines of bacteriology; it was an "antiscientific" attitude that had to be abandoned in the face of sound experimentation. "Some thousands of negative experiments must on the contrary disappear before one single positive experiment, *properly performed.*" Yet, he argued, it was impossible to set up a bacteriological experiment, which could *not* be attributed to faulty technique, that would disprove bacteriological notions of asepsis. "This is why bacteriologists have been led to accept asepsis of healthy tissues of higher animals as a dogma and really cannot act otherwise without letting this domain of science fall into confusion."[57]

Despite his repeated assertions[58] about the power of one properly performed positive experiment over thousands of negative experiments, Portier knew well that the strength of his culturing experiments could not rest solely on the purity and precision of his technique. He offered a series of logical reasons against an interpretation based on contamination:

1. It is always the same organs which give the cultures (in order of frequency, testicle, ovaries, pancreas).
2. These organs gave cultures more frequently when they are taken from vertebrates placed lower in classification, in other words, those in which the *milieu intérieur* of the animal is less independent from the *milieu extérieur.*
3. His results agreed with previous experiments of Victor Galippe and Gabriel Bertrand.
4. The microorganisms from a series of vertebrates were always very similar in the morphology and biochemical properties.[59]

But there was still one more crucial argument against the idea that the organisms Portier cultured were simply the result of bad technique, that is, contamination by infectious parasitic germs. Mitochondria, he asserted, were bacteria, and many competent histologists, including Friedrich Meves, Félix Henneguy, Emanuel Fauré-Fremiet, Claude Regaud, and Alexandre Guilliermond believed they were normal inclusions in the cells of all organisms (with the exception of bacteria). Portier's claim that mitochondria were microbes was fundamental; almost all the theoretical edifice constructed in *Les Symbiotes* rested on it.

Mitochondria Are Bacteria

Portier reviewed the divergent and often conflicting interpretations of the origin and role of mitochondria. In his opinion, the true physiological significance of mitochondria had been suggested by Altmann, and later by Etienne Regaud, who developed his "eclectosome" theory, according to which mitochondria choose and select substances from the surrounding cytoplasm, condensing them and transforming them into diverse products.[60] As Portier saw it, this theory had been proven by Guilliermond, who was able to follow the trajectory of mitochondria in the epidermal cells of the iris flower. According to Guilliermond and his many followers, the elaboration of materials by mitochondria could be direct or indirect. In the latter case, mitochondria transform themselves into leukocytes and result in chloroplasts, amyloplasts, or chromoplasts.

Mitochondria exhibited a striking morphological resemblance to bacteria whether they were in the spherical form resembling micrococci, strung together like streptococci, or in the elongated form of bacteria. Their similarities had been noted by many of those who studied them, in France as elsewhere.[61] Portier asserted that several leading cytologists following Altmann, including Otto Bütschli at the University of Heidelberg and Raphaël Dubois, had suggested that mitochondria were indeed bacteria. But such claims were either treated with contempt or severely criticized by many leading French cytologists.[62] Alexandre Guilliermond, among others, believed that Altmann's equation of bacteria and bioblasts had only done damage to the important recognition that should have been accorded to these structures.[63] In Portier's view, Altmann's critics were all perfectly justified in their criticisms of the speculative nature of his assertions.[64]

Arguments based on morphological appearance hardly represented proof at the best of times, but in cell structures which could be and had been mistaken for artifacts of staining techniques, this was especially dangerous.[65] Portier gathered other evidence scattered in the literature to show the similarities of mitochondria to bacteria. For example, both mitochondria and bacteria were the seat of very remarkable synthetic phenomena. Mitochondria actively multiplied in the cell, not only at the moment of cell division, but throughout the course of their existence. Their manner of dividing was exactly the same as bacteria. As far as Portier was concerned there was only one other proof required to show that mitochondria where symbiotic microorganisms.[66] It consisted in extracting mitochondria from the cell, cultivating them in vitro, and testing the biochemical actions of the cultured microorganisms to verify that they could carry out the same biochemical reactions in vitro that are done in the cell.

Cultivating Mitochondria

Even if mitochondria were symbiotic bacteria, Portier reasoned, it might not be possible to cultivate them. After millions of years of evolution, they may have become so adapted to their intracellular environment that they could not survive outside it. Therefore, he argued that the failure to cultivate mitochondria could never be taken as evidence against their symbiotic nature. The inverse, however, was not true: the ability to cultivate them could be taken as evidence for their symbiotic nature. Here then was an experiment that could only weigh in favor of the symbiotic nature of mitochondria and could never be used against it. But there were mitochondria which could be cultivated in vitro? Portier responded, "Yes, without hesitation."[67]

To make this assertion, however, he had to enlarge the scope of entities that had been generally considered to be mitochondria. If one treated the root nodules of legumes with the strains used for mitochondria, one obtained magnificent preparations of microorganisms which, Portier argued, were essentially identical to mitochondria. During the 1890s, several investigators following the work of Hellriegel and Wilfarth showed that these "mitochondria" could be easily cultured in vitro.[68]

Portier further associated these bacteria with those he had cultivated from inside the cells of the larvae of xylophagous insects. All of them were intracytoplasmic and all of them gave beautiful preparations by the techniques used for staining mitochondria. "I conclude: it is proven that certain mitochondria of plants and invertebrates can be cultivated *in vitro*."[69]

Identifying mitochondria with the microorganisms he had cultured from vertebrate tissues was much more difficult. As one climbed the ladder of living beings, mitochondria lived in a milieu that became more and more isolated from the external environment. They would therefore have become adapted to conditions that were so defined, special, and constant that they would never be able to live in any other conditions.[70] Some of his experiences with culturing microbes fell readily into place with this interpretation. For example, it might explain why he could culture microbes from the tissues of lower vertebrates more easily than he could from mammals: the *milieu intérieur* of the former were less specialized than the latter.

But he had greater difficulty accounting for his ability to cultivate mitochondria, if that is what they were, for mammals. He offered a personal opinion, and suggested that there were two kinds of "mitochondria" present in certain cells. Some were mitochondria that were active in the cell. Others were "new" mitochondria which entered the cell with food; perhaps he was able to cultivate this type. The role Portier assigned to these new mitochondria will be considered momentarily; it is enough to emphasize here that it seemed to him that the physiological state of the cell had an important bearing on his ability to culture such symbiotes. In the great majority of cases, he argued, mitochondria could not be cultivated.

All this, he admitted, was pure conjecture. But the results of the experiments were incontestable. Portier was able to cultivate microorganisms from the tissue of healthy animals which possessed morphological and chemical properties that were identical to those of mitochondria. As he remarked, "We will not therefore confuse the facts with the explanation that I suggest."[71]

The Physiological Role of Symbiotes

Few biological problems were left untouched by Portier's theory. Elaborating on the role of symbiotes in the phenomena of nutrition, he argued that the normal classification which distinguished plants and animals should be abandoned.[72] It was better to divide living beings into *autotrophs* and *heterotrophs*. The former were self-sufficient with regard to nutrition and were represented by bacteria. All other organisms, whether plant or animal, which had autotrophic beings (symbiotes/mitochondria/chloroplasts) in the cytoplasm of their cells were heterotrophs.[73]

There was one fundamental symbiote in Portier's scheme; it was widespread in nature, extremely resistant to physical and chemical agents, and endowed with a very extensive morphological and physiological plasticity. It was this that populated the cells of all organisms.[74] The symbiote was the organ of synthesis, elaborating hydrocarbonated material, fat material, pigments, or albuminous substances. What it elaborated depended on the physical and chemical milieu in which it was placed in the cell. The symbiote was not a free, autonomous entity in the cell. In Portier's view it "(pardon the expression) submits to the cell and is domesticated by it. It is the cell that governs, and exercises control over the symbiote."[75] The nucleated cytoplasm makes the milieu and the symbiote responds and exploits it.

The symbiote, he argued "is the obligatory intermediator between the *milieu extérieur* and the cell; it is a transformer."[76] This view, he reasoned, brought the nutrition of cells in line with digestion by phagocytosis, primitive digestion which persisted in a concealed form. He supposed that once upon a time, there existed (and still may exist today) unicellular beings deprived of symbiotes, reduced to a nucleus and cytoplasm, and which fed on bacteria. Little by little certain of the bacteria adapted themselves to the intracellular milieu and became symbiotes taking the office of intermediaries between the *milieu extérieur* and the cell. When the cells aggregated to form metazoa, the *milieu intérieur* was created and the symbiotes established the link between this milieu and the cell.[77]

If symbiotes were responsible for nutrition, they were also responsible for malnutrition. Symbiotes were constantly dividing; each one was subject to degeneration and their elaborative properties were constantly declining. This, Portier insisted, was not a mere hypothesis; senescence was a generality in nature. In animals, it was essential that symbiotes be renewed, and this had to be carried out quickly and in accordance to metabolic needs. Symbiotes therefore must be carried in food. However, mitochondria/symbionts were not simply replaced by symbionts coming from the *milieu extérieur.* After all, as Portier had argued, the relationship of the mitochondria to the host cell was very intimate due to millions of years of coadaptation: "I do not believe that the symbionts of the food come to *replace* the symbionts of the tissues; instead they would fuse with the old symbionts and *rejuvenescence* results precisely from this fusion."[78]

All the parts of Portier's theory fit together like pieces of a jigsaw puzzle. With "new symbiotes" penetrating periodically with food to rejuvenate the old, he could explain his ability to cultivate mitochondria in vertebrates and at the same time accounted for vitamins. Malnutrition, which the clinicians and physiologists attributed to a lack of an essential vitamin, was due, according to Portier, to a deficiency of symbiotes.[79]

The role of mitochondria/symbiotes was no less essential in fertilization and in ontogenetic development. In fertilization, the young and active symbiotes of the sperm come to rejuvenate, to complete those of the egg: they "produce development." In the case where the egg already possessed young and active symbiotes in sufficient numbers, natural parthenogenesis would occur. If the symbiotes were not sufficiently active, were "quiescent," one could activate the egg by some artificial means. This would account for artificial or experimental parthenogenesis.

Portier supposed that the reproductive rate of symbiotes was normally controlled by the nucleated cytoplasm, but this regulation sometimes was altered and symbiotes reproduced in an uncontrolled manner. Alternatively, there may be an excessive supply of nonadapted foreign symbiotes, which would result in cancer. He explained the fact that x-rays can cause tumors by the unequal sensitivity of nucleus and mitochondria. Other disturbances could be a source of evolutionary change. It was possible that at times, for one reason or another, the *milieu intérieur* would change and prevent the regulation of symbiotes, resulting in sudden or abrupt variations, the creation of a *new species.*

Symbiotes, he speculated, may also have been responsible for the origin of life on Earth. Pasteur had triumphantly demonstrated that the simplest beings originated by filiation, from similar preexisting beings. However, it was neither impossible that spontaneous generation had occurred on the planet in an ancient time, nor that it continues to occur today. All that was certian was that scientists did not know how to recreate the conditions in which spontaneous generation could occur. Others had looked for an alternative solution to the problem. They suggested that life was not born on this planet, that it was inseminated by living germs coming from a different world. This view gained currency toward the end of the nineteenth century, popularized by the Swedish chemist Svante August Arrhenius in his theory of panspermia. These germs would have arrived on the Earth either in the form of cosmic dust or by the intermediacy of meteorites. But there had been serious objections to this suggestion. During their cosmic voyage the germs in question would be exposed to very low temperatures and ultraviolet radiation, if one chose the theory of panspermia, or intense heat, if one chose the meteorite hypothesis.

Portier's work supplied facts which contradicted these objections. As he saw it, they established that interplanetary migration of the symbiote was theoretically possible. But, as interesting as this might be, the origin of life on Earth was too speculative, even for Portier.[80] For him a theory was good scientifically speaking only if it could incite new research and if the edifice did not crumble with the first blow from experimentation.[81] He thus reiterated his central point: "The introduction of the notion of the *symbiote* modifies the sense of the word 'asepsis.' . . . Whether it be cultivatible or not, the presence of a *symbiote* can never be viewed as a contamination."[82] He concluded *Les Symbiotes* by calling for a new kind of bacteriology—"physiological and symbiotic bacteriology"—distinct from that which had been carried out from the point of view of pathology and Pasteurian concepts.[83]

7

The Pasteurization of *Les Symbiotes*

> As soon as it appeared, Dr. Portier's book excited impassioned critiques. For some it was a work that was destined to revolutionize biology, for others it was work made from hastily constructed hypotheses and from experiments which left room for discussion. Admirers and detractors are in agreement, however, in saying that it is rare to find, in a scientific book, originality and boldness of views uniting themselves in a way more favourable to clarity of exposition and seduction of style.
>
> ANNA DRZEWINA, 1919[1]

Les Symbiotes was a synthetic thesis, an interwoven plot addressing the great problems of biology in a new way that defied established doctrines. It was an exercise that trod boldly across disciplinary boundaries, ignoring separate domains of competence. It was a flow of thought that opened up the moment one released microbes from their restricted role as vile germs. But *Les Symbiotes* confronted too many well-established doctrines not to encounter ardent opponents. In the months that followed the publication of *Les Symbiotes,* Portier found himself in heated debates in which his competency as an experimentalist and his authority to discuss and investigate bacteria and cytology were challenged. While some biologists conceded that it was possible that symbiotic bacteria might have a more important role in certain vital functions than had been suspected, others denied that bacteria were other than disease-causing in mammals. But all agreed that Portier pushed his theory too far when he identified symbiotic bacteria with mitochondria and concluded that all living cells contain symbionts in their protoplasm.

Portier's experiences outside the mainstream of French biology, his position at the Institut Océanographique, may have given him the intellectual space to explore the possibilities of symbiosis and to challenge existing doctrines, but these opportunities were not cost-free. He did not have the resources of researchers at the Pasteur Institute, the Sorbonne, or the Muséum d'Histoire Naturelle. He had no great stock collections and, most important, no research assistants. In the midst of the heady debate which ensued, Portier had more immediate and practical concerns—to secure adequate working conditions to continue testing some of his hypotheses and to obtain further results to support the main pillars of his theory.

Making a Revolution—Winning the Election

Portier dedicated *Les Symbiotes* to Prince Albert of Monaco and to comparative physiology, but it was also dedicated to obtaining a position at the Sorbonne. Since 1906, when he had been appointed professor at the Institut Océanographique, he had also worked as an assistant in the laboratory of Albert Dastre, who held the chair of physiology at the Sorbonne. In 1917, when Dastre suddenly disappeared at the front, Portier applied for the vacant position. He wrote *Les Symbiotes* at breakneck speed to ensure that it would appear in time to support his candidacy. The possibility of the Sorbonne position had pushed him to show his theoretical breadth in *Les Symbiotes,* as he did so well. Now that it was written, it was important that it be shown as quickly as possible to those whose opinions mattered.[2]

Portier was especially interested in pursuading Yves Delage, professor of zoology at the Sorbonne. It will be recalled that Delage had attended and indeed was partly responsible for the meeting in Geneva where Merezhkovskii presented his last and fullest account of symbiogenesis. That meeting took place in October 1918, the same month in which *Les Symbiotes* appeared. Delage was familiar with Portier's reports before the publication of *Les Symbiotes,* but he was doubtful of the originality or Portier's claims and the soundness of his experiments. Of course, the idea that mitochondria were bacteria was not new with Portier. Altmann's theory of bioblasts was well-known, and many histologists had made remarks about the morphological similarities of mitochondria and bacteria. In 1917, when Portier published results on symbionts in vertebrates and their role in metabolism, Galippe published his book *Parasitism Normal and Microbiosis,* according to which all living cells contain microorganisms.[3] At first Delage saw Portier's work as being little more than a rehash of Gallippe's ideas of microbiosis. He and his co-worker Marie Goldsmith saw little difference between the two when they reviewed them in *L'Année Biologique* in 1917.[4] Delage also doubted Portier's claims about the physiological role of symbiotes.[5]

Delage expressed his doubts about the originality and plausibility of Portier's claims directly to Portier, who could only hope that his discussion in *Les Symbiotes* would be enough to convince Delage of the error of his initial judgment. Moreover, some new evidence reported from the Pasteur Institute seemed to confirm Portier's claim that microbes could regularly cross the digestive wall barrier in mammals. In December 1918, Pierre Masson and Claude Regaud reported that in rabbits certain bacteria contained in the appendix continually penetrate inside the epithelial cells of the intestine, and from there to the lymphatic tissues, where they are surrounded and digested by macrophages.[6] Prince Albert also had a word with Delage.[7]

Portier's book appeared to have had the desired effect. Delage, it was reported, was moving to his side in the election. Portier wrote to Jules Richard with the good news:

> I have received compliments from various sides which I hope are sincere. As I see it, this exposé of 14 years of work has already changed the opinion of quite a few people who could have a big influence on the election. Delage was clearly opposed to me and did not hide it from the Prince who very kindly intervened for me.
>
> His assistant said to me that after reading my book, he has begun to come over to my side. This is what Delage said to Joublin who saw him yesterday.[8]

Delage and Goldsmith published a lengthy review of *Les Symbiotes* in *L'Année Biologique*. This time they were convinced that "aseptic life, long believed impossible, has been demonstrated" and that "the penetration of microbes across the digestive cell wall barrier was an incontestable fact, and one for which the works of the author [Portier] had provided numerous examples."[9] They were equally impressed by the intellectual coherence of Portier's theory and the evidence he brought to bear showing the morphological and physiological similarities of mitochondria and bacteria, and the suggestive facts relating symbiotes to vitamins.[10] They found his explanation of fertilization to be plausible and looked forward to his further development of the idea that changes in the nature of symbiotes under the influence of diet could lead to large sudden variations and to the origin of new species. They also listed numerous objections to Portier's theory. They argued that the phenomena of senescence and of conjugation, which Portier invoked to explain cases of malnutrition and the action of vitamins, had never been observed in bacteria. In Delage and Goldsmith's opinion, to attribute avitaminosis to a lack of symbiotes was a perfect solution for termites and some other insects but not for other animals. They also found many objections to Portier's explanation for artificial parthenogenesis.

Despite these criticisms, Portier had at least convinced Delage of the merit of his theories and opened his mind to the possibility of symbiosis as a fundamental characteristic of life. Jules Richard wrote to Prince Albert expressing his hope that in the end, the appointment at the Sorbonne would be based on merit, not on politics.[11]

A Showdown with the Pasteurians

There were still other problems, this time coming from the bacteriologists at the Pasteur Institute. Portier, of course, had realized that the bacteriologists would be especially critical. He had provoked them in his book, arguing that for bacteriologists to accept symbiosis as a normal occurrence they would have to abandon the central doctrines that held their practice together. But there was more at stake than a battle over beliefs. As Portier saw it, there was also the question of institutional prestige and professional jealousy: bacteriologists at the illustrious Pasteur Institute would have to accept, or at least give recognition to a new grand theory of bacteria, constructed by an outsider. One of the uninitiated was now going to inform them of the errors of their ways and of the new direction their work ought to take. Suspecting that the bacteriologists at the Pasteur Institute would find this hard to take, Portier wrote to Richard to request funds for further research for the ensuing battle:

> I have received many compliments from diverse perspectives. Apart from a few exceptions, only the Institut Pasteur holds back on the reserve . . . until it becomes really hostile; there are no illusions in this regard. Never will this establishment, equipped with the powerful means of research that you know, forgive one isolated worker for having opened a way that it should have found a long time ago.[12]

Portier's suspicions were not ill-founded; the bacteriologists at the Pasteur Institute had been preparing. The attacks began at the next meeting of the Société de Biologie on 8 February 1918, with comments from two bacteriologists from the Pasteur Institute: Louis Martin and Etienne Marchoux. Neither of them was willing to accept Portier's evidence for the existence of symbiotes in normal healthy tissue of vertebrates. Martin complained that Portier could not find them in all cases.[13] He suggested that the microbes which Portier found in the testicles were simply germs coming from the general circulation which develop preferentially next to or in the testicles. That is, they were due to infection. "In numerous experiments," Martin proclaimed, "we have searched for microorganisms in organs and in general, we have not found any. When we do find them, their presence is readily explicable without the intervention of symbiotes."[14] Marchoux insisted further that bacteriologists had the right to be surprised that Portier was able to cultivate a microbe adapted to intracellular life so easily in normal bacteriological media, when the bacteria of leprosy, which was intracellular, could not be cultivated in vitro because bacteriologists could not make a culture as defined as the cell protoplasm in which it lives. Marchoux challenged Portier to repeat his experiments in collaboration with him:

> In order to shed light on my religion or his, I have already asked M. Portier to repeat his culturing experiments with him. I insist again today that he permit me to witness with him:
>
> 1. That one can take from the testicles of a healthy animal in 50% of cases a microbe which grows in ordinary laboratory media;
> 2. That the microbe is always pure and always the same;
> 3. In 50% of cases where the culture grows, there really exist symbiotic germs in the remaining sterile testicles.
>
> It is up to M. Portier to make this last proof, to indicate the reasons which hinder the culture and the means for making it work in all cases.[15]

Before Portier could reply, another member of the Pasteur Institute, Claude Regaud, demanded further details of his experimental procedures.[16]

Henri Bierry, who had collaborated with Portier on some of the experiments with mammal and birds, responded before him. He found it somewhat curious that the bacteriologists doubted Portier's claims that one could culture microbes from the testicles of animals when similar claims had been published by Galippe and Bertrand, who had not been questioned: "In the face of these precise facts," Bierry concluded, "the *apriori* hypothesis of an accidental contamination cannot be supported."[17]

Portier replied that he had written *Les Symbiotes* in order to produce a theory that seemed to be fertile because it incited numerous studies, and that he himself had made all the objections that had been raised by his critics. Flabbergasted at what he considered to be the personal nature of the discussions, he remarked:

> I fully realized that this theory would encounter resistance, especially among bacteriologists. It seems to me to be useless for us to deliver passionate long verbal duels in order to establish who is right or wrong. It is necessary to search in good faith on the part of each other *for the truth;* there are two different points of view and only experiments carried out over a sufficient time will give us definitive scientific results.
>
> I consider it to be very solidly established that the testicles of diverse healthy animals withdrawn with their surrounding tissue give cultures frequently and outside of all accidental contamination.
>
> I would be very happy to repeat these experiments with the collaboration of MM. Martin and Marchoux.[18]

The agreement to repeat the experiments was formalized by the Société de Biologie, which published the following statement: "Following this exchange of views, the Société de Biologie invited MM. L. Martin and E. Marchoux, on the one side, and P. Portier and H. Bierry on the other, to carry out research together on the following point: Do the testicles and their annexes in a normal state enclose microbes?"[19] The results of this contest were not reported until more than a year later, on 8 May 1920. In the meantime, the controversy in the meetings of the Société de Biologie continued.

Protests and Priority

The encounter with the bacteriologists dealt mainly with only one factual aspect of Portier's theory: whether he actually cultured symbiotes from the tissue of healthy mammals, or whether they were simply infectious germs. It did not concern itself with Portier's interpretation that mitochondria were bacteria. This was left to be savaged by histologists the following month. Their discussions entailed diatribes about what mitochondria *really* were, what they should be called, and who was *really* the first to discover them. In general, the histologists maintained that there were only differences between mitochondria and bacteria.

The first to respond to this aspect of Portier's theory was Claude Regaud. He recognized that mitochondria were universal and fundamental cell organelles, but he protested against the neologism "mitochondria" (thread granules) coined by Benda, named according to their morphology. Instead, Regaud preferred to name them according to their physiological function: *éclectosomes,* as he himself had baptized them because of their elaborative role in electively concentrating substances necessary for the metabolic activity of the cell. He further protested against any significant resemblance between mitochondria and microbes, dismissing Portier's evidence to the contrary. The bacteria described by Portier were resistant to acetic acid and alcohol. However, mitochondria are partially dissolved by them. And although it was true that the methods for staining mitochondria also sometimes stained microbes, Regaud dismissed this as being unimportant since they stained other objects as well. "When all is said in done," he remarked, "apart from a resemblance of form, the exceptional coincidence of similar staining, and the common power of chemical synthesis which appears in all living material, *there are only differences between mitochondria and microbes.*"[20] Regaud also asserted that the reason why Portier confused mitochondria with microbes was because he was not proficient in the techniques for studying cell structures.[21]

Portier countered that although some mitochondria were very sensitive to physical and chemical agents, not all were. If one examined the mitochondria of protozoa, and especially those of sex cells, he argued, one would arrive at a very different conclusion regarding mitochondria. For example, the mitochondria of spermatozoa were much more resistant than those of parenchymal cells. In fact, he reversed the claims of Regaud and asserted that while some mitochondria were very resistant to chemical and physical agents, some authentic and cultivatable microbes contained in the anatomical elements of invertebrates were very sensitive to physical and chemical agents.[22] In making this counterclaim, Portier raised a further issue almost designed to annoy the histologists. He argued that any discussion of the issue at hand required a precise definition of mitochondria. Yet such a definition was absent in all of the classical treatises and original papers which concerned themselves with mitochondria. Therefore, he offered his own definition of mitochondria, one which, by analogy, would embrace the symbiotic bacteria in the root nodules of legumes:

> These are, for me, mitochondria again imperfectly adapted to cell life and capable of being separated from the organism in which they live; they are *promitochondria,* which over the course of centuries will be able to progressively undergo a perfect adaptation that will render them entirely similar to classical mitochondria. For these reasons, therefore, it seems to me that we cannot, *a priori* and in a definitive manner, reject the identification of mitochondria of the cell with symbiotic bacteria.[23]

Regaud was quick to respond that mitochondria were certainly not a vague conception of histologists; they were objects which one could easily define according to the sum of their properties.[24] He defined them as corpuscles of definite form, most often spherical or filamentous, in suspension in the cytoplasm, having the function of elaborating the products of cell activity (secretion products, fat, glycogen starch, pigments, etc.). Regaud insisted that "true mitochondria," unlike bacteria, were extremely fragile, and that what Portier took to be the more resistant mitochondria, for example, the spiral filament of mammalian sperm, were not true mitochondria but rather cell organs of mitochondrial origin.

Departing from his previous response, however, Regaud did not deny physiological similarities between microbes and mitochondria. Instead, he argued that one could not infer from the simultaneous presence of mitochondria and microbes (e.g., the root nodules of legumes) that one could be transformed into the other. Therefore, to suggest that microbes evolved into mitochondria was beyond the scope of experimentation, beyond the boundaries of proper science. It was entering the domain of the unknowable. He further denied any basis for Portier's hypothesis of symbiotes as carriers of vitamins and ridiculed Portier's way of associating phenomena under a grand scheme of symbiosis.[25] Bierry insisted that he and Portier had fully established that certain bacteria were capable of synthesizing vitamins.[26]

This marked the end of the formal dialogue with the Pasteurians. However, their discussions had captured the interest of several other biologists outside of Paris. The criticisms of the relationship between mitochondria and bacteria continued during the next two meetings of the Société de Biologie, 29 March and 5 April 1919.

Critiques from Lyon

Histologists from Lyon came to Paris to offer their opinions or even to declare priority. Alexandre Guilliermond was the first to respond.[27] Unlike the Pasteurians, he began by applauding Portier's theory on the necessity of symbiosis for the most important phenomena in cellular physiology. He was convinced that Portier had demonstrated the presence of symbiotic bacteria in the cytoplasm of certain cells playing a role in assimilation, and that in certain cases he was able to culture them in artificial media. He also agreed that these symbiotes had many similarities to mitochondria. Portier's theory, he remarked, possessed "a very seductive truth": "It is very far from my thinking to contest that symbiosis plays a much more important role than we have admitted until now, because the remarkable work of Dr. Portier carries the proof."[28] Nonetheless, he insisted that symbiotes were far from having the generality that Portier assigned to them.

There was no question for Guilliermond that mitochondria showed a great morphological resemblance to bacteria. They divided like bacteria and were incapable of being formed de novo. However, like Regaud, he emphasized the fragility of mitochondria compared to the "solidity" of bacteria. He claimed that mitochondria were among the most fragile bodies of the cell. This fragility hardly agreed with the extreme resistance of the symbiotic bacteria that Portier had isolated and characterized as being resistant to temperatures of 100°C, able to maintain life in the presence of absolute alcohol and of chloroform. He argued that what Portier took to be more resistant varieties of mitochondria were not "true mitochondria, but plasts differentiated from typical mitochondria."[29] He further insisted that the diverse symbiotic microorganisms described in the literature could never be confused with mitochondria. He asserted that the *bactéroïde* of legumes, fungus, and zoochlorella each possessed a differentiated structure, whereas mitochondria had a homogeneous appearance. It was not difficult, he argued, to distinguish a zoochlorella of hydra from a chloroplast of a plant. Based on these considerations, Portier had provided no demonstration that one could cultivate mitochondria, and Guilliermond concluded that there was no reason to consider mitochondria other than as constituent organelles of the cell cytoplasm.[30]

The criticisms of Etienne Laguesse from Lille differed from the others.[31] He asserted that although mitochondria transmitted themselves from cell to cell, and even from individual to individual at the moment of fertilization, they still did not possess the autonomy that the nucleus did. He believed it to be entirely possible that in certain cells, at least, mitochondria were made by materials emanating from the nucleus into the cytoplasm. He also offered a new name for mitochondria, *ergastidions* ("little workers"), which referred so well to their elaborative role which had become more and more accepted.

The relative consensus over the universality of mitochondria, their ability to reproduce themselves, and their physiological role in plant and animal cells attracted wide attention. In this regard, the Lyonnaise physiologist Raphaël Dubois entered the debates to claim priority for the discovery of mitochondria and their physiological role in the cell.[32] In *Les Symbiotes,* Portier had mentioned Dubois after Altmann, stating that in 1896 Dubois offered more evidence for the existence of these organelles, which he called "vacuolides." Dubois pointed out that in fact it was in 1887 that he first proposed this term to designate the small bodies which captured his attention in his studies of luminous insects, and which he found later in all cells. He believed that they were identical in structure and function to many other elementary granules to which one had given many names around the same time—plastidules, bioblasts, leucoplasts, and so on.[33] As he saw it, that his vacuolides later became known under the neologism mitochondria, named by German histologists, was simply due to xenophilia (love of strangers).[34]

Dubois argued rhetorically that he was not trying to claim priority for discovering that the cell was not the smallest unit of life, but that it was constituted of an agglomeration of many very small organized granulations. This was a very old notion that he traced to the writings of Jakob Henle in 1843. Dubois asserted that for Henle this claim was based neither solely on hypothesis, as it was for other nineteenth-century biologists, "nor on facts simply glanced at or badly observed." He quoted from Henle to show that what he had observed were actually mitochondria, thus undermining the priority of the German Altmann and any others to the discovery of "mitochondria."[35] The idea, of course, was that if Dubois was prepared to graciously recognize Henle's priority, then others in France should do the same for him. He asserted, in no uncertain terms, that all the work conducted since the time in which he wrote simply amounted to confirmations of his own research and what he had been teaching for more than twenty-five years.[36]

In *Les Symbiotes,* Portier had painted Dubois as an ally, stating that he had reluctantly concluded that his vacuolides were symbiotic microorganisms.[37] Dubois dissociated himself from such claims, arguing that though he had entertained the idea at one time, he now rejected it.[38] When studying the phenomena of luminescence in fireflies and certain beetles, which he attributed to vacuolides, he had been struck by analogies between them and certain luminous microorganisms. However, by 1886 he had abandoned the idea on the grounds that it was unreasonable to suppose that symbiotic microorganisms could accomplish a physiological function in an animal. In some cases, he was able to obtain cultures of photobacteria from certain mollusks; but he dismissed this evidence as being due to contamination. He applied this same interpretation to the subsequent work of Pierantoni. He also denied the evidence for symbiotic algae living in certain "green animals"; in this view this was simply "animal chlorophyll." This was the last of the commentaries on Portier's book in *Comptes rendus de la Société de Biologie.*

"Wrong on Every Point"?

Though one might have thought that these kinds of criticisms would hurt Portier's application for the chair at the Sorbonne, the professors there continued to see a significant accomplishment in *Les Symbiotes.* The views of Louis Matruchot, professor of cryptogamic botany at the University of Paris, were representative of this attitude. Matruchot agreed that Portier had gone too far in pushing his theory of symbiotes by likening mitochondria to bacteria. Yet he was well aware of the social stakes in the contest, the battle over disciplinary boundaries, and Portier's candidacy at the Sorbonne:

> M. Portier is above all a physiologist: it is in physiology that he observes and operates, not in bacteriology nor in morphology. And precisely the bacteriologists have the right to ask for greater precision in regard to the techniques employed. . . .
>
> M. Portier works not only in physiology but also in philosophy. He has essentially a synthetic mind; in the relations between things, he sees connecting lines, rather than differences which distinguish them. His book, premature perhaps, and some special circumstances forced him to publish a bit too hastily, is a bold synthesis and, in sum, a real body of doctrine, which extends to practically every domain of biology.[39]

Matruchot was sympathetic to Portier's grand synthesis. As he saw it, the problem was simply too large and complex to be resolved in a few years by a single scientist working in modest conditions. In his view, biologists ought to be grateful to Portier "for having the courage to pose the problem" and for having raised its implications for general biology. He predicted that Portier's "ingenious insights" and efforts to bring so many disparate facts together under a common explanation would not be wasted.

Matruchot's review appeared in September 1919. In the meantime, Portier stood firm in his conviction that mitochondria were symbiotic bacteria. On April 1, he wrote to Jules Richard at the Musée Océanographique about the difficulties with the Pasteurians and outlined the main problem of providing definitive evidence for the symbiotic nature of mitochondria. "Even if one takes all the operative precautions," he wrote to Richard, "when one obtains positive results, it will be very difficult to draw firm conclusions. The Pasteur Institute will urge cries of rage, because it even refuses to accept facts that are much more solidly established."[40] He had not shown that the bacteria he cultured actually were derived from the tissue cells. He had argued only by analogy, by comparing the morphological and physiological characteristics of the organisms he cultured with those of mitochondria. To be convincing, he would have to culture mitochondria directly from the cells which contained them. But in his view, this approach was beyond the scope of existing techniques.

Portier was infuriated by the way the bacteriologists as the Pasteur Institute had set up the contest as a showdown. As he saw it, their hostility was directly associated with his attempts to obtain a better academic position. Though Portier had written *Les Symbiotes* to support his candidature at the Sorbonne, a chair in general and comparative physiology fell vacant at the Muséum d'Histoire Naturelle and he had applied for that as well. But, as he wrote to Richard, already he detected hostility regarding his applications to the museum, hostility which he attributed to the Pasteur Institute:

> They appear to be well enough disposed toward me at the Muséum, but I sense a certain resistance over the past few days. They are beginning to insinuate that if I want to be nominated, I have to resign from the Institut Océanographique. I find this particular pretension to be remarkable on the part of professors who almost always have two or even three positions. The truth is that behind this, the Institut Pasteur is fiercely against me.
>
> They won't forgive me for my book. At every meeting of the Société de Biologie there are new attacks. . . .
>
> Now, I *have* to be *wrong* on *every* point. As they say there, it is necessary that absolutely nothing of my work remains. Also, they multiply the attacks with the aim of tiring me and in the hope that I will admit to defeat and that I will leave the field.
>
> Oh well, they don't know me. In the end, there is nothing more terrible than the "soft in the head." The only great danger is that I am alone without help, with little means of work, and of very mediocre health in face of an enormous and formidably equipped establishment.[41]

If his critics at the Pasteur Institute were hurting his chances of obtaining the position at the Muséum d'Histoire Naturelle, his publisher, Masson, was doing its part as well. By May 1919, *Les Symbiotes* had completely sold out but Masson refused to reprint it. Portier found this attitude to be inexplicable. Yet he found himself in an awkward position since the first edition had not paid for itself, and he himself still owed the publisher the difference.[42] Jules Richard wrote to Prince Albert formally requesting funds to help Portier pay for *Les Symbiotes* and for the publication of an overview of his scientific research to date (*Notice sur les travaux scientifiques*), which he would need for his candidature at the Muséum.[43] Richard also told the prince how happy Portier's adversaries would be to see the end of *Les Symbiotes* and requested that Portier be helped in his struggle with the Pasteurians.

In his exposé of his scientific work for his candidature for the chair at the Muséum d'Histoire Naturelle, Portier was cautious in his appraisal of *Les Symbiotes*. He criticized himself for not always making a clear distinction between fact and interpretation, and he indicated that he would not dogmatically adhere to his theory. "In my thinking," he wrote,

> this small work was only a *theory,* and above all a working hypothesis, as I have remarked many times. But carried away by the exposé of the subject, I have made the mistake of not always making a sufficient distinction between *facts* that I considered to be solidly established and the *interpretations* that I proposed.
>
> However, the discussions that my theory raised, new researches for which the program had been laid out for a long time in my mind, and others born from objections which I have myself made, have already sensibly modified my first opinion on many points; it certainly continues to evolve as experiments bring new results.[44]

Thanks to Richard and the prince, Portier was able to settle the deficit with Masson. The question now was whether to simply reprint *Les Symbiotes* or to publish a substantially new edition, which would correct his mistakes and bring his theory in line with new evidence. On 10 October 1919 Portier wrote to Richard for advice and to inform him of more difficulties at the Muséum d'Histoire Naturelle, where it was insinuated that he did not need the position because he was being "kept" by the prince of Monaco. "P.S. I attach capital importance to my nomination at the Muséum, because it is my scientific liberty, my autonomy and my means of work that are at stake."[45]

Les Mythes des Symbiotes

Portier would soon learn why his publisher, Masson, did not publish his book. In November 1919, Masson released a new book by Auguste Lumière, bluntly entitled *Les Mythes des Symbiotes*. Lumière would have no problem in paying for *Les Mythes des Symbiotes*. A great inventor who, with his brother Louis, had invented the cinematograph—the first motion picture projector—in 1895, he was one of the wealthiest men in France, having created a vast enterprise of paper and photographic products.[46] However, Auguste eventually left the development of cinematography to his brother while he pursued his interest in chemistry and then biology, concentrating more and more on his lifelong interest in medical questions.

In 1896 Auguste Lumière transformed a Lyon hotel into a laboratory for experimental physiology and pharmacopoeia from which many new medications originated, including hypnotics, disinfectants, a hair-dressing formula, and a procedure for embalmment. In 1910, when Auguste Lumière inaugurated his new research laboratory for inventions, some fifteen researchers worked in five sections—chemistry, physics, histology, serology, and analysis—with a scientific library of some thirty thousand volumes. But it is his work in biology that primarily interests us here.[47] Following the discovery of Portier and Richet, Lumière studied anaphylaxis. In the domain of biochemistry, he investigated the colloidal state of life, arguing that the destruction of the colloidal state determined sickness and death. He also studied the therapeutics of tetanus and cicatrization, investigated and theorized on the cause of cancer, and researched vitamins as well as *Les Symbiotes* of Paul Portier.

Lumière began his critique by saluting Portier's daring hypothesis: "Although my conclusions are not in agreement with his views, I beg Dr. Portier to see in this impartial and conscientious study only a dedication rendered to his high intelligence and *esprit d'initiative.*"[48] He acknowledged that "since the immortal work of Pasteur" biologists had come to recognize the growing importance of microorganisms in phenomena of general physiology. There were nitrogen-fixing bacteria in the soil; people used microbes in the agricultural industry and for the preparation of certain foods, fermented drinks, and bread. Microorganisms participate in innumerable analytic or synthetic reactions made at the cost of vital products and of organic wastes, which are modified and reenter in the general cycle of the transformation of matter. Indeed, he assured readers that one could not even conceive of life on this planet without the intervention of microorganisms.[49]

After mentioning this, Lumière retreated to viewing microbes from the perspective of medicine. To suggest that microorganisms were indispensable to cell life, and that all cells enclose them, he declared, was a baseless generalization. First, he rejected the premise underlying Portier's ideas: that there would be a perfect and stable intracellular equilibrium without a struggle of elements constituting the symbiotic association. He asserted that there were only a few exceptional cases in which such an equilibrium established itself without struggle between the cell and the "parasite," one destroying the other. But even then, such an equilibrium was certainly not stable.[50]

Lumière repeatedly emphasized the vile nature and aggression of microorganisms, claiming that it was ridiculous to believe that microbes could normally penetrate into our tissue to rejuvenate mitochondria.[51] Nonetheless, he acknowledged that most microbes were not noxious and that certain saprophytes, *Bacillus subtilis,* for example, could circulate inside an organism and survive in liver, spleen, and bone marrow for several months after being injected into an animal.[52] Lumière claimed that he and his collaborators confirmed this; they were able to culture bacteria from various animal organs.[53] They further examined whether certain kinds of nonpathogenic bacteria could penetrate into tissue by natural means: they placed frogs in water containing a culture of bacteria and left them there for eight hours. Out of eighteen attempts at culturing, they obtained only one culture of microbe. But Lumière discounted it on the grounds that it was *impossible* to rule out contamination during the procedure.[54] In his view, the symbiotes Portier had cultured were simply saprophytic spores having accidentally passed over the usual barrier which protects the *milieu intérieur* of the animal.[55]

Lumière not only denied the unversality of symbiosis; he also denied the evidence for the universal existence of mitochondria in plant and animal cells.[56] "In the great majority of cases," he wrote "the *symbiote,* so clearly visible in cultures '*in vitro*' has therefore passed to the state of a phantom."[57] He also insisted that metabolic processes did not rely on the presence of *life within life,* but rather on chemical substances. He was not alone in taking this attitude. Anna Drzewina prefaced her remarks about Portier's book with the same kind of comment. "It was not very long ago that scientists who studied the problems of heredity localized them in the chromosomes of the nucleus. When mitochondria were discovered one transferred the seat of heredity there."[58] Lumière applied this same criticism to the claim that vitamins were carried in symbiotic bacteria.[59]

Lumière interpreted his work as a wholesale refutation of Portier's claims. But all Portier saw were confirmations. One could indeed cultivate microbes from the tissue of healthy animals. The only differences he saw were interpretations. Yet he had also been modifying his interpretations as new results came in over the previous year. His aim was to publish a second modified emotion of *Les Symbiotes.* As he wrote to Jules Richard in November 1919, "Have you seen the book by Lumière? In sum, it confirms all that I have said from the point of view of *facts.* Only the interpretation differs. In a few months, perhaps before, I will send them *un coup* of 420 pages which will make them reflect."[60]

By this time Richard's enthusiasm for Portier's defense of *Les Symbiotes* had dwindled. Was Portier being too insistent? Was his obstinancy going to hurt his chances of obtaining the position at the museum or even the position at the Sorbonne? Was he ever going to convince the bacteriologists and histologists in France of the universality of symbiotes, of their relations to vitamins, their role in development, fertilization, heredity, and variation, with the poor techniques then available, with so little help, with the relatively meager facilities Portier possessed at the Institut Océanographique? Should he give up, go back to business as usual, be less provocative, do something smaller but solid, bow out and make his mea culpa as he said he would do if he were wrong when he first sent his manuscript to Richard two years earlier? Portier remained confident in the course he was following, as he wrote to Richard the following week:

As for *Les Symbiotes,* be patient and don't be defeatist. All the material facts that I have advanced have been confirmed by Lumière. Let the questions evolve and you will see that I know perfectly well where I am going. The only important thing for me is that I possess some means of work. The Prince has promised me work, even *offered* it, in a manner which touched me infinitely. I am forced to accept his offer with much gratitude.[61]

A second edition of *Les Symbiotes* was never published. The following month Portier learned of the results of the nomination for the chair in physiology and the Muséum d'Histoire Naturelle. He lost the election. In December 1919, Jules Tissot, assistant at the Muséum, was offered the position. In May 1920, Portier was appointed professor without chair at the Sorbonne.[62] That same month the results of the contest between Portier and Bierry on one side and Marchoux and Martin on the other were published in short note in the *Comptes rendus de la Société de Biologie.* They were inconclusive. It was "impossible to affirm" that microbes were normally present in the testicles and their annexes:

1. The passage of pieces of organs of an animal in the culturing media is always difficult to achieve with constant asepsis. It is one of the most delicate operations in bacteriology.
2. One does not generally obtain cultures from healthy organs when one uses the pulp of testes collected on average from an unravelled Pasteur tube.
3. One may find microbes in the testes in some conditions and some proportions when working with whole organs or large fragments.[63]

Bacteriophage and "Microlichen"

Though a professor at the Sorbonne, Portier was permitted to keep his position at the Institut Océanographique. But his work on symbiosis dwindled as his research conditions strengthened. On 22 June 1922, Prince Albert I of Monaco died. At that time, Portier headed an active laboratory with eight students working on the physiological role of vitamins and various other problems of general physiology. None worked on symbiosis.[64] In 1923 a new Chair of Comparative Physiology was established at the Sorbonne for Portier.

In 1919 and 1920 Maurice Caullery (1868–1958) taught a course on symbiosis at the Sorbonne. It resulted in his book *Parasitism and Symbiosis* (1922), which was translated into English thirty years later. Caullery reiterated the criticisms against the idea that symbiosis was a primordial characteristic of cell life, as suggested by Portier and others.[65] In fact, Caullery seemed reluctant to write about bacteria. He said very little about bacteria, except to assert "that most bacteria pathogenic or not are parasites." "In point of fact, bacteriology and the problems it raises, have, on account of their theoretical and practical importance, their place in other works; in this one, therefore, they have been left to one side."[66]

Ironically, by 1923, just when the study of symbiosis was receding from the foreground of Portier's research, it seemed to him that it was gaining acceptance and attracting great attention at the Pasteur Institute. The French Canadian Félix d'Herelle had claimed that bacteria were also susceptible to infection and were hosts to ultramicroscopic, filter-passing agents, which he named *bacteriophages* (eaters of bacteria).[67] Portier saw them as nothing other than symbiotes. When in 1923 he was asked to help provide a scientist at the Pasteur Institute with information regarding marine bacteria for studies of bacteriophages, he remarked, "I am amused from my side that those who had been so much angered by my little book are taken to study this question which did not exist before."[68]

Research on bacteriophages raised considerable controversy in many countries during the 1920s. Even d'Herelle's priority in the discovery was challenged. Two years before d'Herelle's first report, in 1915, the English bacteriologist Frederick Twort published an article on a filterable agent that destroyed bacteria.[69] D'Herelle himself denied Twort's priority in the discovery.[70] But the central question to researchers was the nature of this agent. Was it an infectious virus or an enzyme produced by the bacterium? Twort left these questions unresolved.

D'Herelle insisted that the destructive agents were viruses, "an ultramicrobe, a filtrable being endowed with the functions of assimilation and of reproduction."[71] He thought that they might be used as important agents in the fight against pathogenic bacteria and that they play an important role in natural immunity in addition to that played by phagocytosis. However, his interest in the relations between bacteriophage and host was much broader and included the nature of symbiosis and what he considered to be its primary role in evolutionary change. Indeed, studies of symbiosis, especially those of Noël Bernard, provided d'Herelle with crucial arguments for the ultramicrobial nature of the bacteriophage and its behavior.

Soon after his first report on bacteriophages, d'Herelle observed that not all bacteria were destroyed by them. He asserted that "mixed cultures of bacteriophage and bacteria" could be subcultured indefinitely.[72] In 1919, he described several profound transformations in the morphology and physiological properties of bacteria which had resisted the action of the bacteriophage *Bacteriophagum intestinale,* which he claimed normally existed in the intestinal tracts of animals. He amplified these studies in his first text, *The Bacteriophage: Its Rôle in Immunity* (1922).[73] The virulence of the bacteriophage and the resistance of the bacteria, he argued, "are not fixed, but are essentially variables, being enhanced or attenuated according to the inherited properties of each of the two germs, and according to the circumstances of the moment which favor the one or the other of the two antagonists. These two phenomena dominate the pathogenesis and the pathology of infectious disease."[74]

In his second book, *The Bacteriophage and its Behavior* (1926), d'Herelle explicitly referred to the perpetuation of mixed cultures in terms of a symbiosis, an equilibrium due to a continual selection to the more virulent bacteriophage and the bacteria of greatest resistance. Referring to the studies of Noël Bernard, he argued that "the respective behaviour of the bacterium and the bacteriophage is exactly that of the seed of the orchid and of the fungus."[75] The balance between the resistance of host and the virulence of the parasite persists as long as the immediate environmental conditions remain unchanged. He also pointed out how *Paramecium bursaria* and *Hydra viridis* are "normally parasitized" by a minute green alga, *Zoöchlorella vulgaris*.[76] Symbiosis played no minor role in evolution in d'Herelle's view. In his words, "symbiosis is in large measure responsible for evolution."[77] Like Bernard before him, he was reluctant to use the term symbiosis "for it dates from an epoch when it was thought that the communal existence of the two beings necessarily involved mutual benefit. We now know that symbiosis is always a struggle, which as a rule, ends in the elimination of one of the two contestants."[78]

But d'Herelle suspected that the relationship between bacteria and bacteriophages did not always lead to such an elimination. "Sometimes a perfect symbiosis develops and it is then permanent."[79] Indeed, he referred to such mixed cultures as "microlichens." The controversy around the dual nature of lichens offered the historical precedent against those who would object to the perpetuation of such an intimate relationship between two organisms and the construction of dual organisms:

> A mixed culture, comprising bacterium and bacteriophage, is in reality a culture of microscopic lichen. A bacterium in symbiosis with a bacteriophage is a "microlichen." And it would not be strange . . . if certain bacterial species may be recognized only when in the form of "microlichens," that is, as mixed cultures.[80]

D'Herelle maintained his interest in the transformations of bacteria produced by bacteriophages. He published a paper on "Bacterial Mutations" in 1931, again referring to the work of Noël Bernard, and asserting: "Botany offers many examples of such sudden mutations brought about by symbiosis, and even in man such modifications occur, as examplified by the changes in the skeleton observed in congenital syphilis."[81]

The role of the bacteriophage in heredity did not rise to the center of genetic discussions until some twenty years later (see Chapter 11). Many biologists continued to reject the existence of virus-harboring, or lysogenic, strains of bacteria. D'Herelle had said nothing of Portier's claims that mitochondria were symbolic bacteria. However, as will be discussed in the next chapters, d'Herelle's assertion in 1926 that "symbiosis is in large measure responsible for evolution" was elaborated by the American biologist Ivan E. Wallin into a bacterial conception of mitochondria. By that time, Portier's commitment to the idea that mitochondria were symbionts was dwindling.

"The Thankless Task of Precursor"

Portier continued to publish on subjects ranging from comparative and general physiology to marine biology to the biology of insects until 1952, when he was eighty-six years old. He received many prizes and honors: in 1923 he was named Officer de la Légion d'honneur; in 1929 he was elected a member of the Académie de Medicine. He was not elected to the Académie des Sciences until his retirement in 1936 when he was Professor Honoraire. In preparation for this nomination he assembled another account of his scientific work and included in it a discussion of his book *Les Symbiotes.* This time, he retracted any claims for identifying mitochondria with bacteria. In his view, the most serious objection that others had raised against the bacterial nature of mitochondria was the fragility of mitochondria to physical agents, as opposed to the resistance of bacteria. However, Portier continued to assert that mitochondria were not as fragile as others had supposed.

> But, there is a stronger objection, it seems to me, against likening mitochondria to bacteria; it is that in certain cells (the nodules of legumes . . .) bacteria and mitochondria coexist and *that one can distinguish them.* I believe to have been the first to give this refutation, and I have, moreover, never defended this theory; I do not believe that today it can be sustained.[82]

This alone was certainly not crucial evidence against the theory that mitochondria were symbiotic bacteria. In fact, seventeen years earlier, Portier himself had recognized that mitochondria and symbiotic bacteria had coexisted in the root nodules of legumes. He had referred to the symbiotic bacteria as *promitochondria,* arguing that they would ultimately evolve into organelles similar to actual mitochondria.

As to the facts he advanced, Portier saw none which had been disconfirmed experimentally. There were indeed microorganisms which existed normally and constantly in the tissue of animals. And microorganisms did play an important role in the digestive process of xylophagous insects.[83] Moreover, when he had published *Les Symbiotes,* reports dealing with symbiosis were few and scattered in the literature, and they referred above all to plants. But by the mid-1930s, Portier claimed the question of symbiosis had taken on considerable importance, and he pointed to numerous cases of symbiosis described in all orders of animals: protozoa, sponges, ceolenterates, turbelarians, echinoderms, rotifers, and mollusks, as well as many insects. Many of these studies had been brought together by Paul Buchner in *Tier und Pflanze in Symbiose* (1930), a volume of almost a thousand pages. From the extensive work described by Buchner it was clear that very important physiological processes could be accomplished only thanks to the intervention of microorganisms living in the tissues of animals. Portier considered his role in the new burst of work on symbiosis to be that of a catalyst: "If, in 1918, when presenting the importance of this beautiful question of symbiosis, I formulated some interpretations and hypotheses, I am pleased to state that I have had some merit in assuming the thankless but useful task of a precursor.[84]

8

Les Symbiotes Revisited

Their universal presence in the cell coupled with the known properties of bacteria, appear to indicate that mitochondria represent the end adjustment of a fundamental biologic process. The establishment of intimate microsymbiotic complexes has been designated "Symbionticism" by the author. Symbionticism, then, is proposed as the fundamental factor or the cardinal principle involved in the origin of species.

IVAN E. WALLIN, 1927[1]

The evidence for microorganisms playing an essential role in the physiology of organisms accumulated through the 1920s. Portier's hypothesis stimulated interest in symbiosis in France, and the work of Paul Buchner (1886–1978) and his students in Germany provided more evidence of symbiosis in insects. Buchner is recognized today by some as the "master and founder of systematic symbiosis research"[2] and especially for his compiled great work, *Endosymbiose der Tiere mit pflanzlichen Mikroorganismen,* the fourth edition of which was published in English, *Endosymbiosis of Animals with Plant Microorganisms* (1965).

Buchner enrolled at the University of Munich in 1907 with the intention of becoming a botanist, but in his third semester he attended the lectures of the celebrated zoologist Richard Hertwig. At that time, Richard Goldschmidt, who later became a prominent geneticist, was a lecturer and Hertwig's assistant. Buchner chose Goldschmidt as his principal supervisor. Although his dissertation was on "The Accessory Chromosome in Spermatogenesis and Ovogenesis in Orthoptera," his interest in genetics and chromosomes was short-lived. After completing his doctoral work, he began a year-long stay at the Zoological Station in Naples. There he fell under the spell of the blue gulf and its islands, one of which, Ischia, he was later to adopt as his home. He also fell under the spell of symbiosis.

While in Naples, Buchner attended the lectures of Umberto Pierantoni on symbiosis in sap-sucking insects such as cicadas and aphids. The pseudovitellus was recognized as a special primitive gland organ in the gut tube populated with intracellular symbiotic yeast. As Buchner later recalled, "With the publication of these studies it seemed as though a blindfold had been removed from the eyes. Numerous examples of symbiosis in the Homoptera were described in rapid succession, and numerous other important insect families or smaller systematic entities were recognized as symbiont barriers."[3] He returned to the University of Munich as assistant to Hertwig in 1910, and a few years later he took over Goldschmidt's vacated position as curator.

During these years, Buchner studied intracellular symbionts in sap-sucking insects and their hereditary transmission through the egg cytoplasm. He demonstrated that arthropods (plant lice, scale insects, plant bugs, cicadas, etc.) that suck plant juices have symbiotic yeast contained in special gland cells, which he called "mycetocytes" of primitive organs which he called "mycetomes." Like Portier before him, he argued that the yeast cells produced enzymes which act as ferments for digesting food for the insect. These studies provided a broad platform for Buchner's studies of hereditary symbiosis in other insects, including ants and cockroaches. He also followed the studies of Pierantoni on the role of bacteria in luminescence. In 1920, he turned to study insects that suck the blood of vertebrates. Lice, bedbugs, ticks, mites, tsetse flies, mosquitos, and fleas all contained symbiotic bacteria; their relatives that suck invertebrate blood had none. Buchner assumed therefore that the symbionts helped break down red blood corpuscles in vertebrates.

By 1921, when Buchner published his first book, *Tier und Pflanze in intrazellularer Symbiose,*[4] he recognized that symbiosis was far from a haphazard occurrence; it seemed to him to have rules: partnerships between animals and green algae were confined (with exception of sloths) to transparent and water-living creatures; it was only there that the food producing activities of the green cells could be turned to advantage. One could find them in protozoa, sponges, and other coelenterates, flatworms, and certain snails. When the "plant partners" were algae, they helped in food manufacture; when they were fungi or bacteria, they helped in food utilization.

Although Buchner believed such symbioses to be of wide occurrence, he opposed any suggestion that symbiosis was a primordial characteristic of cell life. In his view, symbiosis was "always a supplementary device, enhancing the vital possibilities of the host animals in a multiplicity of ways."[5] In 1921 he wrote a critique of Portier's work and throughout his career he dissociated his work from what he considered to "be a bold hypothesis based on extravagant concepts." Nevertheless, he did recognize that symbiosis was a significant source of evolutionary novelty. As he wrote, "If we refuse to create a completely new principle of cell structure through intracellular symbiosis, it does not necessarily follow that intimate symbiosis may not be the stimulus for the development of new animal forms. We have previously indicated such possibilities in the insects."[6]

In 1923 when the Cambridge parasitologist George Nuttall, president of the physiology section of the British Association for the Advancement of Science, gave an address entitled "Symbiosis in Animals and Plants," he introduced it as a subject that at least potentially had a broad biological interest. However, he complained that the literature relating to symbiosis was "largely foreign, somewhat scattered and relatively inaccessible."[7] Nuttall offered a brief statement on the controversy in France over Portier's *Les Symbiotes,* how "like many exploded hypotheses" it had served a useful purpose by provoking discussion and interest in the subject of symbiosis. Nuttall emphasized that symbiosis was by no means "so rare a phenomenon as was formerly supposed," that symbionts could become, in some cases, permanent inhabitants that could be transmitted from host to host hereditarily. In his view, the increasing number of infective diseases of animals and plants which had been traced to viruses suggested that there may even exist ultramicroscopic symbionts:

> It is evident that we are on the threshold of further discoveries, and that a wide field of fruitful research is open to those who enter upon it. In closing, it seems fitting to express the hope that British workers may take a more active part in the elucidation of the interesting biological problems that lie before us in the study of symbiosis and the allied subject of parasitism.[8]

Similar opinions about the scope and significance of symbiosis were maintained by K. F. Meyer, a renowned plague expert who held a dual appointment at Zurich and in the Medical School of the University of California at San Francisco. In 1925 he wrote an extensive overview on symbiosis including his own work on symbiotic bacteria in terrestrial prosobranch snails. He mentioned how the books of Portier and Buchner had aroused new interest in "a relatively virgin field of investigation."

Meyer emphasized how difficult it was to distinguish between a symbiont and a cell product and how various bodies such as the bacteria in the root nodules of legumes were believed to be modifications of cytoplasm before they were shown to be able to grow in artificial media and shown to be able to infect young plants grown on sterile soil.[9] It was obvious to him that "the 'symbiotic' processes were concerned in the evolution of certain species of animals and plants."[10] He found it impossible to imagine how, for example, such structures as mycetocytes and mycetosomes in various kinds of insects could have developed without altering the physiology, morphology, and consequently the habits of the host. He also suggested that knowledge of symbiosis would be useful for understanding such problems in pathology as local immunity and tumor formation due to parasites.

The next year, the protozoologist L. R. Cleveland at the Department of Tropical Medicine, Harvard University Medical School, promoted the experimental study of symbiosis as "a most fruitful field of investigation awaiting an opening."[11] Cleveland himself became well-known for his studies of termite symbiosis. The intestines of termites were teeming with a menagerie of intestinal protozoa which he proclaimed weighed almost as much as the insects themselves. "Everyone who has observed them has marvelled at their abundance, interesting structures, peculiar adaptation and mode of living, and has wondered how on earth so many of them live in a single termite, what they do, that is how they live and what relation they have to the termites in which they live, move, and have their being."[12]

By the mid-1920s, approximately fifteen hundred species of termites had been described and grouped into four families. Every species of four out of the five families of termites contained symbiotic protozoa; Cleveland experimentally demonstrated that these protozoa were necessary for digesting wood. The presence of microsymbionts in termites raised many questions: What were these termite-inhabiting protozoa? Has the evolution of symbiotic protozoa been going on longer in termites than in other animals, or did it mean that protozoa have found a much more favorable environment in termites than in other animals? Does a such a symbiotic association encourage evolution?[13] The only family of termites that did not contain them did not feed solely on wood and when they ate it was much more decayed than that which the protozoa-harboring families eat. Yet even some of the termites which did not harbor protozoa formed partnerships with fungi. It seemed that they gave over a certain portion of their nests to the growth and cultivation of the fungus, which they eat along with the soil and cellulose the fungus has acted upon. Some researchers had observed what actually appeared to be the deliberate removal of certain parts of the so-called fungus garden during prolonged droughts followed by a week or so of rain to propitious places outside the nest.[14]

Rediscovery in the United States

Despite these promotions for symbiosis as a field ready for exploitation, few English-speaking biologists took up the experimental study of symbiosis. Examples of intracellular symbiosis, as widespread as some believed them to be, were confined to some plants and invertebrate animals. The idea that all plant and animal cells had evolved and were maintained by symbiotic associations was either regarded as fantastic and improbable or ignored by virtually all biologists of the 1920s and 1930s. There was, however, one notable exception, Ivan E. Wallin (1883–1969), of the Department of Anatomy at the University of Colorado School of Medicine.

Like Portier before him, Wallin constructed a far-reaching theory of the bacterial nature of mitochondria and the role of microsymbiosis in development and heredity. His research on the bacterial nature of mitochondria culminated with his book *Symbionticism and the Origin of Species* (1927). Wallin had completed his doctoral research at New York University in 1915, on tissue development, differentiation, and morphology in the lamprey *Ammocetes.*[15] His subsequent interest in mitochondria and symbiosis in evolution reflected the spread of such studies from Europe to the United States. The extensive cytological work and theorizing on the physiological role of mitochondria by French and German biologists was introduced into the United States by Edmund Vincent Cowdry.

Cowdry had studied mitochondria at the Peking Union Medical College of the Rockefeller Institute and in 1918 published an extensive review of the mitochondria literature.[16] He discussed Regaud's eclectosome theory, according to which mitochondria served as organs of elaboration, and Guilliermond's evidence for the origin of chloroplasts from mitochondria. Cowdry also mentioned the similarity of mitochondria to bacteria; but like many others, he regarded Altmann's theory of bioblasts to be an expression of an exaggerated yet understandable mistake, as well as one of the reason for the neglect of these granules by cytologists.

Written before Portier's book appeared, Cowdry's main argument was not against the bacterial nature of mitochondria. This came later, when Wallin began to write on the subject. In the meantime, Cowdry's polemics were directed against nucleocentric views of heredity.[17] Advocates for the view that mitochondria possess genetic continuity, he remarked, "do not say or even hint (as adherents of the chromosome hypothesis have not refrained from doing, in the case of chromosomes) that the mitochondria constitute in any sense of the term the sole basis of heredity."[18] He was not exaggerating. In 1926, the leader of the *Drosophila* school, T. H. Morgan (who was later awarded a Nobel Prize), wrote, "In a word the cytoplasm may be ignored genetically."[19] Cowdry cautioned biologists against making a hasty conclusion: "Those who make the conservative statement that mitochondria play some part in heredity occupy just as secure a position as those, on the other hand, who claim that chromatin [the chromosome] is the sole heredity carrier."[20] While this might have been true in some abstract intellectual sense, in any institutional sense, the latter viewpoint occupied a much more secure position.

Genetics in the United States evolved into a discipline in its own right, with its own journals and societies, doctrine, aims, theories, and well-defined techniques. While geneticists organized themselves into a discipline, studies of mitochondria had been carried out largely on an individual basis, by scattered researchers in various countries, often isolated by war, language, priority disputes, and institutional competition. This was also true of those who pointed to symbiosis as an evolutionary strategy; they had persistently operated in conflict and competition, constantly undermining or ignoring one another. Wallin followed this same pattern in his treatment of Portier's work when he introduced it to American biologists.

Translation and Transformation

Published at the end of World War I, and within a few months sold out in France, *Les Symbiotes* was not a highly visible text in English-speaking countries, to say the least. A brief review by an anonymous author appeared in the British journal *Nature;* it concluded:

> Prof. Portier is good-humoured enough to quote the paradox that a theory is not of value unless it can be demonstrated false. We have no hesitation in prophesying that his theory will attain that value—which is just what he would have said himself a few years ago. We are bound to admit that the author is a downright good sportsman.[21]

No review ever appeared in the equivalent American journal, *Science.* Wallin himself first learned of it in 1922 when he submitted for press the first in a series of seven papers "On the Nature of Mitochondria."[22] Not having read *Les Symbiotes,* Wallin was quick to protect the independence of his own discovery and the possibility of his own priority. He stated that he had for some time been developing the idea that mitochondria were bacteria, but he "was not ready to state this hypothesis until [he] had collected more evidence in its support."[23]

Portier's book was something that Wallin could not ignore. He hired a translator and the following year introduced *Les Symbiotes* to American readers in a paper entitled "A Critical Analysis of Portier's 'Les Symbiotes.' "[24] However, Wallin's account of Portier's experiments and theorizing was less than precise. His misrepresentations began with the assertion that Portier believed that mitochondria, or symbiotes, "were constantly entering and leaving the host organism under normal conditions."[25] Actually, Portier said nothing about symbiotes constantly leaving the host organism. As to their constantly entering the host, Wallin gave the following account:

> After entering the host they first occupy an intercellular position and later take up an intracellular relationship. In the intercellular relationship the organisms retain their hardy properties and are easily cultivated in culture media. After the "symbiotes" have taken up the intracellular position they lose their hardy qualities, develop fragility and are quite incapable of cultivation in culture.[26]

In this passage Wallin is referring to Portier's suggestion that new mitochondria, or symbiotes, are ingested by the intestinal epithelial cells. However, Portier never suggested that these symbiotes actually became mitochondria. To the contrary, he emphatically stated that the symbiotic bacteria entering from outside the organism did *not* become adapted to intracellular life. Instead, he suggested that they fuse with, so as to rejuvenate, mitochondria, which had become adapted to intracellular life over millions of years. This notion, of course, fit well with his claim that symbiotic bacteria found in foods were carrying essential vitamins. Wallin had mistaken part of Portier's argument for the whole, and he repeatedly emphasized wrongly throughout the 1920s that Portier believed that mitochondria *originated* from food and entered the organisms by way of the alimentary tract. He compared what he took to be Portier's belief with the "correct" view that although symbionts could enter through the alimentary tract, "the perpetuation of such a new symbiotic complex is by way of the germ cell."[27] Again, Portier fully agreed with this claim.

Wallin's third major misinterpretation of Portier's *Les Symbiotes* was that Portier had attempted to cultivate mitochondria: "The chief basis for Portier's conclusions rests on his results from attempts to grow mitochondria in artificial culture media."[28] This misattribution has been perpetuated in the English-language literature since Wallin first introduced it in 1923. Suffice it to say, Portier had never attempted to grow mitochondria in artificial media. He had attempted to cultivate microbes from the tissues of healthy animals and had argued by analogy that these symbiotes were like mitochondria. He had also argued by analogy that the bacteria in root nodules of legumes were mitochondria, or rather "promitochondria."

Wallin misrepresented Portier's work again when he asserted that "the premises for Portier's researches rest on a statement attributed to Louis Pasteur in which he maintained that the tissues of a healthy animal are free from bacteria."[29] Wallin insisted that Pasteur's statement concerning the aseptic tissues of healthy animals was erroneous. He referred to work done in the United States during the 1920s that led some to suggest that bacteria were present in 80 percent of apparently healthy animals. His own results were consistent with this suggestion. He claimed that he was able to get "luxuriant bacterial growth on the ordinary agar culture media from lymph nodes of apparently healthy young rabbits and kittens."[30]

Since bacteria were normally present in the tissues of healthy animals, Wallin reasoned, that it was obvious that the fundamental basis for Portier's experiments was faulty.[31] This would be true if Portier actually had attempted to cultivate mitochondria. However, his fundamental aim was to demonstrate the very phenomenon that Wallin cited others for showing—that microorganisms were present in the tissue of healthy animals—from invertebrates to vertebrates. This was the very issue which had led him into confrontation with the bacteriologists at the Pasteur Institute. Yet in his subsequent writings Wallin persistently referred to the false premises of Portier's experiments.[32] Portier never responded; there is no evidence to suggest that he ever read Wallin's work.

Wallin never reviewed Porter's theories concerning the physiological role of mitochondria in cell life, fertilization, development, and variation. Instead, he dismissed them as being based on speculation. What one needed, he proclaimed, were hard facts on the nature of mitochondria.[33] Yet his own theory was no less speculative. Despite his omissions, false interpretations, and criticisms of Portier's work, the evidence Wallin marshalled for the bacterial nature for mitochondria, the rhetoric he used, and the theories he advanced were strikingly similar to Portier's. Indeed, Wallin made his work appear not only independent but almost as if it were spontaneously generated.

Local Debates in a Global Context

Like Portier, Wallin traced the reluctance of biologists to accept the view that mitochondria were bacteria to fundamentally opposed perspectives of the nature of bacteria and of cell physiology. The claim that symbiosis was a fundamental characteristic of life was swimming against a tide of the all-pervasiveness of individual life struggle. The claim that the symbionts in question were bacteria only seemed to make matters worse. Bacteria had always received bad press, constructed as the enemy of all humans, the source of disease, misery, and death. The disease-causing attributes of bacteria stood in direct contrast to the life-giving functions of mitochondria, as the research interests of pathologists stood in virtual conflict with those who studied symbiosis. These issues became explicit in the polemics of Wallin's adversary Edmund Cowdry, who in 1921 had moved from Peking to the Rockefeller Institute for Medical Research in New York.

Cowdry's main interest in mitochondria was their use in studies of cellular pathology. He claimed that they offered pathologists a cytoplasmic criterion of cell activity as well as a nuclear one. However, following the appearance of Wallin's papers, Cowdry received many queries about the possible symbiotic nature of mitochondria—a question which in his view took the mitochondria problem off the main track. As he saw it, the proposed bacterial nature of mitochondria was not only wrong but inasmuch as it crossed over the specialist boundaries between cytologist and bacteriologist, the suggestion could only lead to confusion since few biologists were competent to evaluate it. Cowdry laid out the stakes in a paper written with the virologist Peter Olitsky and published in the *Journal of Experimental Medicine* in 1922:

> Although the evidence presented by Altmann, Portier and Wallin is not convincing, certain inquiries which we have received indicate that the discussion provoked may easily lead investigators from the main problem. . . . Particularly is this true since the problem lies, as it were, between the sciences of cytology and bacteriology, so that but few investigators can be familiar with both sides of the question.[34]

They further asserted that verification of the bacterial nature of mitochondria would have profound and deleterious effects on both physiology and pathology.[35] Upholding the views of F. G. Hopkins, the founder of biochemistry at Cambridge, they maintained that in physiology it was not "profitable to attempt to distinguish between living and non-living constituents": life was considered to be a function of the metabolic activities of the cell as a whole. They claimed that the admission that all cells contained living entities would "force a complete adjustment in our ideas of the organization of living matter and in the application to it of the laws of physics and chemistry."[36]

The effect of such a theory would be no less serious on pathology since, in their view, it would obscure concepts of disease and the means for detecting their causes. The disease known as rickettsia was a case in point. In 1909 an American bacteriologist, Howard Ricketts, had observed minute, bacterialike bodies in the intestines of certain ticks which transmit Rocky Mountain spotted fever. Similar bodies were seen in 1916 by Rocha-Lima in lice taken from cases of typhus fever. Believing these to be living organisms and the cause of the disease, he named them *Rickettsia prowazekii* honoring Ricketts and Prowazek, both of whom died of typhus fever during the course of their investigations. *Rickettsia* had been cultivated on special blood media, but they could not be grown like ordinary bacteria.

In Cowdry and Olitsky's view, the identification of mitochondria as bacteria would turn bacteriology on its head in terms of the criteria to be employed to identify such pathogenic microorganisms.[37] To ward off such potential confusions, they emphasized what they saw as the crucial differences in the physical and morphological properties of mitochondria and bacteria. They also argued that a definition of bacteria was required before the theories of Altmann, Portier, and Wallin could be considered acceptable. They quoted a passage from Park and Williams's text, *Pathogenic Microörganisms,* in which disease was a characteristic of bacteria.[38]

Cowdry and Olitsky's article was followed by an unsigned editorial in the *Journal of the American Medical Association,* "Are Mitochondria Identical with Bacteria?" It reiterated the fear of the negative effects that a bacterial interpretation of mitochondria would have on current ideas of the organization of living matter and pathology. The editorial concluded:

> Amid this uncertainty, reassurance comes from the new studies of Cowdry and Olitsky at the Rockefeller Institute for Medical Research. A direct comparison which they have made between mitochondria and bacteria in the living condition reveals microchemical differences which can be attributed only to fundamental dissimilarities in chemical constitution. The spores and capsules characteristic of bacteria have not been noted in mitochondria; motility due to flagellar action has not been observed. In fermentation or disease, bacteria tend to increase, whereas the mitochondria in cells tend to decrease, and there is no reason to believe that mitochondria possess the power of independent and characteristic growth apart from cells. The theory of "symbiotes" thus remains without convincing support, while the possible importance of the widely present mitochondria in cell physiology still looms large.[39]

Wallin published a response to Cowdry and Olitsky's criticisms the next year, refuting each of their claims, and dismissing their remark that investigations of the bacterial nature of mitochondria threatened to lead biologists from the main problem.[40] First, he challenged their staining evidence which distinguished bacteria from mitochondria. They had not been able to stain mitochondria with the Giemsa stain and concluded that this pointed to a fundamental difference between mitochondria and bacteria. Wallin asserted that he was able to stain mitochondria with the Giemsa stain in smear preparations of dog fetus tissue dried in the air and followed by absolute alcohol fixation.[41] He rejected their claim that a bacterial conception of mitochondria would "leave pathologists in some doubt regarding the criteria to identify microorganisms the nature of which remains obscure" as being simply "farfetched." More to the point, it was irrelevant in regard to the truth or falsity of the bacterial nature of mitochondria. As Wallin put it, "The known nature, relationships and reactions of Rickettsia bodies leave no basis for any possible confusion with mitochondria, but if there were occasion for the suggested confusion the writer is convinced that *no criterion* is preferable to a *false criterion* in the search for truth."[42]

With regard to a definition of bacteria, Wallin argued that the passage quoted by Cowdry and Olitsky did not constitute a definition of bacteria and was apparently not so intended by Park and Williams.[43] Biologists usually distinguished bacteria morphologically from unicellular plants and animals by the absence of an organized nucleus. But Wallin insisted that one could never construct an adequate definition of bacteria based simply on a catalogue of their morphological and physiological characteristics. One also needed to situate them in an evolutionary and ecological context.[44] He pointed out that only a small percentage of bacteria were pathogenic: using Park and Williams's book on disease and making disease an essential characteristic of bacteria was absurd.

There were many strains of bacteria which had an intimate relationship to other forms of life but were not necessarily pathogenic in their reactions. Wallin pointed to the *Bacillus coli,* which is nourished by the food in the intestines and in turn renders the food more easily digestible by the host organism. But aside from a few forms such as the bacteria in the root nodules of legumes, he argued, there simply had been very little study of the nature these important life relationships.[45] Cowdry and Olitsky never responded. In 1923, Cowdry left New York to study rickettsia in South Africa.

For Wallin, "the ultimate proof" of the bacterial nature of mitochondria demanded "the demonstration of mitochondria growing independently outside of the living cell."[46] The tissue to be used for inoculation of the culture medium had to be as free from contamination as was possible to find in nature. He reported his results in 1924. He planted bits of the tissue from the liver from embryonic, fetal, and newborn rabbits, guinea pigs, and dogs in a number of more commonly used media. The results were negative. However, he could obtain cultures on a human blood medium, but problems in procuring sufficient quantities of human blood to make the medium on a large scale and difficulties with sterilization of the media made this approach inviable. He therefore devised special media, and when bits of liver tissue from the fetal and newborn rabbits were planted in these media, he obtained coccoid organisms that could be fixed in a mitochondria fixative.

Wallin exclaimed that is was merely "a convenient matter" to explain his results away "with the dogmatic cry of 'contamination.' The danger of contamination always lurks in the background of bacteriologic technique and it appears that this 'bug-a-boo' has often served to misdirect the search away from the true course."[47] He assured his readers that all instruments and vessels had been thoroughly sterilized. The fetus or newborn rabbit after decapitation was saturated with 95 percent alcohol. The instruments used in opening the abdomen and removing the liver were always sterilized in the flame immediately before use. The liver was quickly removed to a sterile petri dish, cut into pieces by inserting a sterile scalpel under the lid. Pieces of the liver tissue were then planted in the media by the usual bacteriological technique.

Based on this evidence, and that which he had submitted in previous papers, Wallin declared in 1924 that he had proven that "*mitochondria are, in reality, bacterial organisms, symbiotically combined with the tissues of higher organisms.*"[48] Nonetheless, and despite all his precautions, searches for alternative explanations, and reassurances, by 1927, when he wrote *Symbionticism and the Origin of Species,* Wallin's experimental demonstrations for the bacterial nature of mitochondria had been neither refuted nor confirmed. There were no demands for witnesses or counterdemonstrations.

Theory Mongering in the United States

To make the symbiotic nature of mitochondria plausible one had to do more than accumulate physical and chemical comparisons of mitochondria with bacteria and experimental evidence of their relative autonomy. Wallin delved ever further into the realm of theory. Like Portier before him, he elaborated a conceptual niche whereby the presence of bacteria in all cells could address contemporary problems in biology. However, theory mongering was considered bad form in biology in the United States. Biology had differentiated into more or less semiautonomous specialties, each proclaiming that theorizing per se should be put aside in favor of collecting "facts."

In the third edition of *The Cell in Development and Heredity* (1925), E. B. Wilson altered the opinion expressed in previous editions of *The Cell,* allowing more for the possibility that cell components originated as symbionts in the remote past. Based on the new evidence of invisible genes and invisible viruses minute enough to pass through a filter, he suggested that many of the diverse bodies of the cytoplasm—including centrioles, plastids, and mitochondria—might be derived from diverse self-reproducing entities that lie beyond the power of the microscope in the apparently structureless ground substance.[49] He had also become more impressed with evidence for symbiotic algae in Protozoa and lower Metazoa. In the first edition of *The Cell* he had doubted that the chlorophyll bodies were symbiotic algae and favored the view that they were "animal chlorophyll." By the mid-1920s the tide of opinion favored the symbiotic interpretation.[50]

This evidence, as well as the consensus regarding the symbiotic nature of lichens, made Wilson a little more cautious in dismissing symbiotic theories of cell components offhand as "totally baseless." Nonetheless, his attitude to such theories was all too apparent when he referred to Merezhkovskii's ideas as "entertaining fantasy" and as "flights of the imagination."[51] He did not mention Portier, but he did refer to Wallin's claim that mitochondria were symbiotic bacteria. Again Wilson was warming to the possibility: "To many, no doubt, such speculations may appear too fantastic for present mention in polite biological society; nevertheless it is within the range of possibility that they may some day call for some serious consideration."[52]

When addressing the issue of theory construction, Wallin was careful to assure his readers that his claims for the bacterial nature of mitochondria were not constructed solely by speculation, but that theory was required for beginning exploratory work.[53] "Digressing into the realm of theory," Wallin examined the compatibility, the attractions, and the repulsions with respect to existing concepts in biology. "The conception that all living cells are loaded with bacteria," he wrote, "is so startling and antagonistc to our orthodox notions of the cell, that without further analysis and reflection the conception appears absurd."[54]

The proposed bacterial nature of mitochondria contradicted orthodox *concepts* of the cell. But Wallin insisted that it harmonized perfectly with the known *facts* about cell activity. In this regard, he distinguished his views from those of Altmann, who had argued that only the bioblasts were alive, and that the cytoplasm of cells in which they lived was lifeless. Wallin claimed that there was abundant evidence to prove that the cytoplasm, but itself, possessed properties that are characteristic of living matter. He insisted that if Altmann had limited his theory to a consideration of the bioblast alone, "then the burden of proof would have fallen on his adversaries."[55]

Wallin maintained that the principles of symbiosis were also consistent with the concept of Metazoa and Metaphyta in terms of the mutual interdependence of cells. Such a relationship was also present among bacteria and had been designated as symbiosis by bacteriologists.[56] The bacterial conception of mitochondria only extended the "cell dependence principle" to include an intimate dependence of all highly organized cells on the activities of simple cells. The *Bacillus radicicola* in the root nodules of legumes was "incontrovertible testimony" that such a functional aand morphological relationship existed. Moreover, the dependence of higher life on bacteria in a different sense had been embodied in the concepts of all biologists, as exemplified in the "nitrogen cycle." It was therefore well within "the domain of logic to extend this well-established principle of dependence of higher life on bacteria to include the more intimate dependence of life processes on bacteria."[57]

As Wallin saw it, a bacterial conception of mitochondria was not antagonistic to established factors in evolutionary theory either. He asserted that it was "in harmony with principles of biological behavior as exemplified in 'the struggle for existence' resulting in symbiosis."[58] In 1923 he proposed the terms "symbionticism" and "prototaxis" as "two fundamental biological principles." "Prototaxis" was proposed "to signify the innate tendency of one organism or cell to react in a definite manner to another organism or cell."[59] Biologists were familiar with "chemotaxis": the orientation of an animal in relation to a particular chemical, the response being negative (moving away) or positive (moving toward)—the movement of a wasp toward an attractive odor such as beer; the behavior of bacteria moving toward sugars; and when white blood cells, leukocytes, attacked invading bacteria in the blood stream, it was assumed that the invading microorganism emitted some chemical substance to which the leukocyte responded. Prototaxis involved chemotaxis, but it also included surface tension, light, temperature, moisture, and perhaps electrical potential.

"Symbionticism" signified "the expression of a positive prototaxis which results in an absolute symbiosis." "Such a symbiosis is of an intracellular type, and one of the symbionts is always a bacterial organism."[60] Wallin was not consistent in his use of this term in its restricted reference to intracellular bacterial symbionts. He recognized the symbiosis of algae and fungi in higher lichens as well as algae living inside animal cells to be absolute symbioses, and he also recognized d'Herelle's bacteriophages as indicating that bacteria themselves could play the role of "relative or host symbiont."[61]

Wallin's theory of symbionticism also fit well with such accounts of the origin of life as that proposed by Henry Fairfield Osborn in *The Origin and Evolution of Life* (1917). Osborn's conception included a chemical evolution leading to first organisms of a bacterial nature, able to subsist on inorganic material. It was not difficult, Wallin argued, to conceive of the role that the primordial chlorophyll-bearing organisms played in the modus operandi of early evolution. Concerned with the production of starch, they assumed the role of food factories preparing the way for the evolution of "higher life."[62] The simplest organisms containing chlorophyll were bacteria or bacterialike organisms, blue-green algae. Thus he cast bacteria in the role of the primordial organisms from which all higher forms of life have sprung.

Wallin was not arguing, of course, that all "higher forms" had simply descended or evolved from one or a few forms of bacteria, but rather that higher organism were constructed by and still composed of bacteria. Like Portier before him, he appealed to comparative physiology to support his claims. Bacteriologists demonstrated that bacteria synthesize chemicals which had identical reactions to enzymes produced in plants and animals. Some products of cellular activity, for example, insulin, were apparently of identical nature in plants, fish, and humans. It had also been claimed that certain bacteria produce or contain insulin. All this could be easily explained in light of the bacterial conception of mitochondria:

> If we should ask the question: How have the specific functions, such as, for example, the synthesis of adrenalin, been acquired by animals in phylogenetic development, it is obvious that we are asking a question, the correct answer to which would furnish the clue to the fundamental factor in the origin of species.
>
> It appears reasonable and logical to suppose that such a function as adrenalin synthesis in the animal organism was acquired when a microorganism having the property of synthesizing adrenalin became symbiotically combined with the animal in question.[63]

There were established precedents for this hypothesis. The symbiosis literature contained a number of examples in which symbionts produce enzymes that were essential to the host organism. In some cases the host responds to the "infection" with a development of a special organ to harbor the bacterial symbiont. Wallin interpreted such responses as being a "direct record of the manner in which organs have developed in phylogeny."

The Inheritance of Acquired Characteristics

In *Symbionticism and the Origin of Species* Wallin systematically addressed problems of paleontologists, developmental biologists, and geneticists, as well as the belief among many biologists during the 1920s that natural selection acting on minute variations could not account for the origin of new species. Buchner's *Tier und Pflanze in intracellularer Symbiose* revealed a wealth of evidence which supported and emphasized the significance of microsymbiosis in the origin of species. Various symbiotic relationships produced permanent morphological and physiological changes in the host.

In the hydroid Portuguese man-of war, the structure of cells that contained algal symbionts was different from other cells. In *Adamsia diaphana,* a hexacoral, the cells of the septa that contain symbionts were morphologically distinct from the remaining septal cells. Buchner had shown that algal symbionts were carried in special cells in a large number of animals. Morphological reactions in the host were especially shown in the turbellarian worms *Convoluta roscoffensis* and *C. paradoxa,* in which the excretory system fails to develop, unlike other turbellarians. The agenesis of an entire organ system was somehow associated with the presence of algal symbionts; the loss of activity of an organ could become hereditary. The presence of special cells containing bacterial symbionts in cockroaches was another example of a morphologic response on the part of a host to the presence of microsymbionts; and the microsymbionts were transmitted from one generation to another in the germ cell.

Among plants, the formation of root nodules in conjuction with the invasion of microorganisms represented a classic illustration of a morphological response in the host. Botanists and bacteriologists had also demonstrated the presence of nodules containing bacteria on the roots of a number of plants other than legumes. A number of plants contained extracellular symbiotic bacteria in the leaves and other parts. In some cases it had been shown that microsymbionts were transmitted in the seeds from one generation to the next. But the relationship in all lichens was particularly instructive. Since the various species of lichen exhibited different degrees of symbiotic relationships, it was difficult to doubt that the origin of all lichens depended upon the establishment of a symbiotic relationship.[64]

Based on his demonstration of the bacterial nature of mitochondria and the responses of organisms to microbial invasions, Wallin advanced his theory that symbionticism was a fundamental factor in the origin of species.[65] Bacteria were the "building stones," the "primordial stuff," from which all higher organisms had been constructed and modified. "Just as reproduction ensures the perpetuation of existing species," he asserted, "symbionticism ensures the origin of new species."[66]

There were three major features in Wallin's evolutionary theory, each governed by a fundamental principle. Symbionticism was the fundamental principle controlling the origin of species. Natural selection was the chief principle governing the diversity of life on Earth; it controlled the retention and destruction of formed species. An unknown principle was responsible for giving evolution direction or progress toward an ever more complex end—to gradual perfection leading to humans.[67] Wallin's views of the limitations of Darwinian theory were representative of the general attitude of many biologists of the 1920s.[68] While some maintained that species emerged suddenly through macromutations, others continued to embrace neo-Lamarckian conceptions of the inheritance of acquired characteristics. The expression "acquired characteristics" took on a different meaning when associated with symbiosis. Normally, it conveyed the idea of modifications occurring in an individual as the result of such environmental influences as temperature, light, moisture, and chemical and physical agents. Other acquired characteristics were physiological in nature, produced by use and disuse. One also included characteristics acquired artificially, for example, through dehorning of cattle, cutting of plumage, and castration. Wallin emphasized that the characteristics associated with symbiosis were of an entirely different nature and relationship to the organism: the acquisition of a microorganism and the morphological response of the host constituted acquired characteristics. Hereditary symbiosis, whereby an acquired symbiont eventually was transmitted in the germs usually through the egg cell, he argued, represented a case of the inheritance of acquired characteristics.

Wallin accepted the possibility that environmental influences acting on bacteria might produce permanent modifications. But he doubted that environmental influences alone could ever have been responsible for the evolution of plants and animals and the origin of species.[69] He found it difficult to understand how environmental influences alone could, for example, bring about changes in a group of cells or microorganisms which would result in the production of an entirely new chemical product. He argued by analogy: just as two or more chemical were essential in the production of a new, more complex molecule, so too was symbiosis necessary for the production of new species.[70]

Thus Wallin offered an account of the origin of true cells, how these led to multicellular plants and animals, and how different species originated, all according to symbionticism. He began by constructing a logical story of the stages in evolution by which one-celled plants and animals might have originated from the bacteria. A first stage involved the emergence of primordial strains of bacteria which derived their nutrition directly from inorganic chemical compounds. A second stage involved the production of *variations* in the primordial bacterial strains when chlorophyll-bearing bacteria, or blue-green algae, arose.

The next stage witnessed the introduction of "true cells"—those which had an organized nucleus contained within a nuclear membrane—"the cellular animals and plants."[71] These emerged when different strains of bacteria fused into a single mass to form an organism of an order higher than that of either component. With successive and compounding fusions the bacterial chromatin increased, and perhaps became morphologically modified with each fusion, ultimately forming the nucleus surrounded by the cytoplasm. Subsequent fusions between the nucleated cell and various strains of bacteria resulted in myriads of species of one-celled organisms, some of which may be in existence today. One could also "recognize in these plants and animals the 'partially fused' microörganisms (mitochondria) that appear to have been and undoubtedly still are the fundamental factor in the production of new forms."[72]

Algae deviated from protozoa in attracting blue-green algal symbionts, which Wallin, like Merezhkovskii, believed to be closely related to bacteria.[73] Various cell structures such as Golgi bodies as well as the locomotor structures, cilia, and flagella were the products of bacterial–mitochondrial symbiosis:

> The ciliate and flagellate protozoa, apparently, acquired their special locomotor structures through symbiosis with ciliate and flagellate bacteria. We have previously mentioned that some investigators have claimed that cilia formation in the cells of higher organisms is associated with mitochondria.[74]

In Wallin's scheme, mitochondria were not organisms that had entered some primitive microbes millions or billions of years ago and, once constructing the prototype, coevolved with plant and animal cells or persisted as relics from the dawn of organic evolution. They were constantly being added to the germ plasm in the course of the origin of species and they produced various cell structures (in accordance with Regaud's eclectosome theory). Addressing the problem of cellular differentiation during ontogeny, Wallin suggested that the mitochondrial population in the germ cell represented "pleomorphisms" of a great number of strains of acquired bacteria. In the chemical environment of the egg cytoplasm the various strains were resolved into a common morphological type. But in later ontogenetic differentiation of the tissues, when new environments are produced, the mitochondria differentiate hand in hand with the tissue differentiations.[75] Therefore, inasmuch as ontogeny involved mitochondrial differentiation, the different cell lines that occurred in ontogeny were the results of various symbiotic events in phylogenetic evolution. In short, ontogeny recapitulated a symbiotic phylogeny.

The evidence suggesting that mitochondria were involved in cellular differentiation and that microsymbiosis could lead to new kinds of cells and organs converged with and supported the idea that mitochondria were bacterial symbionts. If the evolutionary origin of the specialized cell depended upon the microsymbiont, then the hereditary or ontogenetic origin of the specialized cell must also be dependent upon the same factor: mitochondria. As Wallin saw it, "The question of cellular differentiation resolved itself into determining the nature of the modifications which a microsymbiont undergoes in absolute symbiosis."[76]

The inheritance of acquired symbiotic bacteria was no less important for genetics than it was for embryology. Wallin offered it as a solution to one of the major genetic problems of the second decade of the century: the origin of new genes. Geneticists maintained that changes in genes were the major source of evolutionary variation; the process by which such changes occurred was the subject of considerable debate. The leader of the Mendelian movement in England, William Bateson, had postulated that perhaps gene mutations were due to the loss of some factor. This idea was based on the observation that most gene mutations were recessive and seemed to result in the loss of a character that the dominant had. If one believed that gene mutations and rearrangements were the sole source of evolutionary change, then Bateson's idea implied that evolution "from amoeba to man" had proceeded by an unpacking and reorganization of an original gene complex. Not surprisingly, this idea was severely criticized and dismissed by American geneticists, who argued that it led to absurd conclusions.[77]

Wallin noticed the problem. He argued that the genetic concepts which presupposed that evolution would have to proceed by unpacking of an originally complex molecule were only justified on the genetic evidence that had accumulated. They failed to account for increased complexity of structure and function due to microsymbiosis.[78] "It is logical to assume that there have been ever-increasing additions to organisms during organic evolution. These 'additions' must have come from the *outside,* and represent true accretions which are responsible for the origin of species."[79] Insisting that "no *fundamentally new* character is produced by the rearrangement or loss by genes," Wallin suggested that part or all of the chromatin of microsymbionts would be given up to the nucleus of the host cell and germ plasm when symbiosis develops, leaving the remains of the microsymbiont in the cytoplasm.[80]

As innovative as this suggestion might appear to be, Wallin was certainly not the first to mention the possibility of new genes coming in from outside the organism. In fact, although it went unnoticed by his critics (indeed to the present day), William Bateson himself had postulated that while recessive varieties result from the loss of a genetic constituent, new dominant varieties originated from infection. In his *Problems of Genetics* (1913), he wrote:

> If we could conceive of an organism like one of those to which disease may be due becoming actually incorporated with the system of its host, so as to form a constituent of its germ-cells and to take part in the symmetry of their divisions, we should have something analogous to the case of a species which acquired a new factor and emits a dominant variety.[81]

Although Bateson had no evidence for such genes coming from the outside, he was insistent:

> Somehow or other, therefore, we must recognize that dominant factors do arise. Whether they are created by internal change, or whether, as seems to me not wholly beyond possibility, they obtain entrance from without, there is no evidence to show. If they were proven to enter from without, like pathogenic organisms, we should have to account for the extraordinary fact that they are distributed with fair constancy to half the gametes of the heterozygote.[82]

Wallin's theory of the inheritance of acquired symbiotic bacteria would also be useful in addressing some of the results of paleontologists which indicated that "cataclysms" had occurred resulting in the rather sudden appearance of new species in certain geologic areas.[83] For a possible clue to how this might have occurred, he pointed to studies of the behavior of pandemic theoretical niche in biology, the theory that symbiosis was a major source of evolutionary innovation went virtually unheeded during the 1930s and 1940s. On the one hand, his vision of bacteria as the building blocks of life continued to be overshadowed by the war waged against microbes. The year before Wallin published his book, Paul de Kruif's *Microbe Hunters* celebrated the triumphs of those "death fighters" who hunted the microbial menace—Pasteur, Koch, and Ehrlich, who had declared that "we must learn to shoot microbes with magic bullets."[88] On the other hand, the role of symbiosis in the origin of new genes was overshadowed by new techniques for producing gene mutations.

The same year that *Symbionticism and the Origin of Species* appeared, H. J. Muller published his first report that the use of heavy doses of x-rays could increase the frequency of gene mutations in *Drosophila* by some fifteen hundred times. The following year, L. J. Stadler published the results of similar studies on barley. The types of phenotypic changes that occurred were the same as those known to occur spontaneously.[89] The artificial production of mutations gave genetics a new lease on life. Gene mutations, not bacteria, were the "building blocks of life." In his famous essay "The Genes as the Basis of Life" (1926), Muller argued that the ability of genes to mutate and to reproduce themselves in new form conferred on these cell elements the properties of the building blocks required by evolution.[90]

Symbiosis as a major source of evolutionary innovation was also overlooked by the architects of the "evolutionary synthesis" of the 1930s and 1940s, R. A. Fisher, Sewall Wright, J. B. S. Haldane, Julian Huxley, Theodosius Dobzhansky, and later by Ernst Mayr and G. Ledyard Stebbins. The evolutionary synthesis, forged out of a consensus among Mendelian geneticists, population geneticists, and some field naturalists, was based on a "rediscovery" of neo-Darwinism—that natural selection acted on small genetic variations within populations.[91] Gene mutations and recombinations provided the essential fuel for evolutionary processes. The slightest change which favored the survival of an individual would never go unnoticed by the keen eye of natural selection. Evolution would be gradual, based on an accumulation of almost imperceptible changes.

That new organisms could be synthesized from two or more different species—that corals, most termites, all lichens, many protists, many invertebrates, sponges, coelenterates, turbellarians, rotifers, mollusks, annelids, and so on—were symbiotic complexes was virtually ignored in all the major debates over evolution. Symbiosis was regarded as an obscure topic, at best a curiosity, nothing worthy of generalization. Even among disciplines such as embryology and ecology, which were hardly embraced by the so-called evolutionary synthesis, there was little or no room for symbiosis as source of evolutionary change.

Falling in the Cracks

While American geneticists focused on chromosomal genes and claimed them to be governing elements, walled off from the rest of the cell and dictating its diseases that periodically have swept the world such as the great influen: pandemic of 1889–1890 and the second catastrophic wave in the fall an winter of 1918–1919, which killed more than twenty-one million people.[8 These phenomena, he argued, illustrated the possibilities of bacterial behavior for evolution.

This was a creative suggestion that followed logically when one turned medical perspectives of germs into biological ones. Although numerous attempts had been made to isolate a distinctive organism associated with these pandemics, the consensus of opinion among pathologists appeared to be that the organism associated with the disease was normally harmless and present at all times.[85] This implied that some unknown extrinsic factors were responsible for modifications in the microorganism and/or in large numbers of humans. In Wallin's view, however, it was premature and futile to attempt to analyze the factors involved in the development of symbiotic complexes:

> These factors are the forces that influence protoplasmic behavior such as temperature, light, moisture, pressure, electricity, and possibly gravity, solar and lunar forces. In connection with disease, and also symbiosis, we do not know whether the host or the microorganism, or both, are modified by these extrinsic influences. Perhaps, when we some day have acquired some more specific data on meteorological influences on living matter, we may then better understand the behavior of living organisms.[86]

A Failed Revolution

No matter how profound and revolutionary it appeared to him, Wallin's theory of symbionticism moved no one. He investigated it no further. A brief review of his book appeared in *Nature,* in which "symbionticism" was criticized for the "horrid" word that it was. The reviewer, J. Brontë Gatenby, was not even willing to admit the *possibility* that mitochondria were bacteria. Indeed, on Gatenby's own account, no experimental evidence could ever convince him of the bacterial nature of mitochondria:

> Dr. Wallin will find that the field of spermatogenesis alone will provide facts which make his position untenable. We believe that the case he makes for the growth of mitochondria on nutrient media is unsatisfactory, though even if the evidence had been complete, this would not convince the reviewer that mitochondria are necessarily bacteria. . . . The case Dr. Wallin tries to make for a similarity of chemical constitution between bacteria and mitochondria is completely unconvincing.[87]

Wallin's theory of symbionticism and the origin of species was going to win or lose not simply on the basis of its own intrinsic merits—plausibility, logical consistency, and coherence—but on its ability to harmonize with the general aims, methods, and doctrines of the major biological disciplines. Despite his attempt to make his theory compatible with existing ones, and to construct a activities, the slogan of many embryologists was "the organism as a whole." Since the turn of the century there had been a reaction against the idea of localizing the seat life in any particular cell constituent, against the speculative theories of those nineteenth-century biologists who had proposed the existence of self-reproducing submicroscopic entities, and against any theory of life which assumed life.[92] Thus many opposed the notion of dictatorial genes as well as any theory which reduced organismic organization to competition and cooperation among self-producing morphological units.

Throughout the first half of the twentieth century, leading embryologists repeatedly argued that the constitution of the single cell, or complex organism, could not be explained as a symbiotic union of elementary organisms. The American embryologist E. G. Conklin recognized that various parts of the cell possessed the ability to assimilate, grow, and divide. "But it has not been demonstrated," he wrote in 1940, "that any of the visible parts of cells, much less these ultramicroscopic parts, are capable of showing these properties of life *when they are completely isolated from all other parts. In short the entire cell is still the ultimate vital unit capable of showing all these properties of life.*"[93]

Conklin ridiculed any attempt to localize the basis of life in any one kind of substance or cellular constituent. All such proposals involved "the error that the properties of the whole are to be found in the constituent parts" and were suggestive "of ancient views as to the localization of certain emotions in the heart, others in the kidneys or bowels, or Descartes' proposal that the pineal gland is the seat of the soul."[94] Conklin declared, "*Life is not found in atoms or molecules or genes as such, but in organization; not in symbiosis but in synthesis.*" After mentioning the ideas of Altmann, Wallin, Boveri, Watasé, and Minchin, he remarked, "Unfortunately for all such speculations there is no evidence that any of these cell constituents are capable of independent existence."[95]

The importance of symbionts inherited through the egg was equally overlooked by those experimentalists who attempted to established definitive genetic evidence for cytoplasmic inheritance. Between the two world wars several German botanists set out to challenge what they called the "nuclear monopoly" of the cell and detailed various cases of non-Mendelian, cytoplasmic inheritance in plants.[96] They designated the cytoplasmic genetic components of the cell the "Plasmon" in contrast to the genome of the nucleus. Cases of the transmission of infectious agents were often discussed in reference to the evidence for cytoplasmic inheritance. However, the onus on investigators of cytoplasmic heredity was to demonstrate that cytoplasmic constituents constituted an essential part of the genetic system of all organisms.

In such debates the inheritance of characters due to the transmission of symbionts was a source of error and criticism on both sides. It was often referred to as "pseudo-inheritance." It was something that had to be avoided for a definitive demonstration of the universality of cytoplasmic genetic systems.[97] Indeed, some of the evidence for cytoplasmic inheritance was criticized by defenders of the "nuclear monopoly" on the basis that the inherited morphological changes in question might be due to the transmission of symbionts or parasites. As the American geneticist E. M. East wrote in 1934 when warning against "false" evidence for the inheritance of acquired characteristics and non-Mendelian inheritance:

> Thus there are several types of phenomena where there is direct transfer of alien matter capable of producing morphological changes. It is not to be supposed that modern biologists will cite such instances, when recognized, as examples of heredity. But since an earlier generation of students used them, before their cause was discovered, to support the inheritance of acquired characteristics, it is well to be cautious in citing similar, though less obvious, cases as being illustrations of non-Mendelian heredity.[98]

One also has to consider that symbiosis as a source of evolutionary innovation was left out of "the evolutionary synthesis" inasmuch as cooperation in ecology was ignored. The British plant physiologist F. G. Gregory addressed this issue explicitly in 1951 when arguing that one should widen the concept of the organism to include symbiotic complexes:

> The "struggle for existence" presupposes antagonism between organisms whether or not they belong to the same or diverse species. On the other hand, the question remains whether associated species tend to provide for each other a favourable environment. The analysis of the relations between organisms has been dominated by the notion of "competition" or "struggle," and the converse notion of "cooperation" has in consequence been disregarded. . . . The data of ecology serve as a challenge to this view of the predominant role of "struggle."[99]

We now turn to such discussions of cooperation, the underlying dynamics of symbiotic associations, and what counts as an organism. We shall see that those ecologists to whom Gregory refers, those who emphasized cooperation, challenged interpretations of cases which had often been considered classic examples of mutualistic symbiosis and denied that they led to associations that were as well integrated as parts of an organism. Indeed, during the first half of the century, the consensus among biologists of every disciplinary background was that stable intimate symbiotic associations were exceptions, and, furthermore, they could rarely be considered to be of mutual benefit.

9

Verbal Phantoms

> The double danger of research into this type of phenomenon lies, on the one hand, in bringing to them preconceived ideas of too subjective a nature, bordering on an illusory anthropomorphism, and on the other hand, trying to reduce complex facts to simple elementary reactions.
>
> MAURICE CAULLERY, 1952[1]

Throughout the first half of the twentieth century, symbiosis continued to be discussed and understood in terms of laws and principles about conflict and cooperation in nature. Empirical investigations of symbiotic associations remained scarce, and there was a plurality of views about how they originated and were maintained and about their place in natural history. In fact, there was still no consensus over what the term symbiosis actually meant or whether it had any legitimate place in scientific discourse.

A Committee on Terminology

Since the late nineteenth century, two major conflicting definitions of symbiosis coexisted side by side. For some, it was a general term to be applied to all cases where different kinds of organisms live together, that is, to parasitic, commensalistic as well as mutualistic relationships. Others used it to refer exclusively to relationships between different organisms which live in close and mutually advantageous association. No matter how the term was used, Anton de Bary was erroneously cited both for the source of the term and for its meaning; often he was listed as supporting the view that "symbiosis" signified relations that were mutually beneficial to the associates.

These conflicting definitions were brought to the attention of the American Society of Parasitologists in 1933, and a Committee on Terminology was appointed to investigate the problem. In its first report, published in 1937, it concluded that de Bary had often been "misinterpreted": he had actually used *symbiosis* as a general term and he included specifically all degrees of parasitism, commensalism, and mutualism. But the almost "casual way" in which the term was introduced in the 1879 pamphlet made the committee question whether it had actually been used previously.[2] Although the committee could find no previous reference to the term, A. B. Frank had indeed used it a year before de Bary. Therefore, two myths had been perpetuated: de Bary had neither originated the term nor restricted it to mutualism.

Still others terms were problematic. Should one use *symbiote* or *symbiont?* And which associate was the symbiont, which was host? The committee consulted a member of the Department of Classics at Harvard University, who informed them that *symbiote* is derived from the Greek *sumbiotes,* meaning "one who lives with," "companion," "partner," whereas *symbiont* had no Greek original. Though the philologist preferred *symbiote,* the unconvinced committee maintained that it was simply "a matter of taste and usage rather than of correctness." The committee took a similar attitude toward the terms symbiont and host. Symbiont could be properly used for either member, though it had become custom to refer to the smaller as the *symbiont* and the larger as the *host,* in conformity with parasitological language.

The definition of the term symbiosis, however, could not be left to whimsy. Still the committee refused to make any formal recommendations except to suggest that the term be defined whenever it was used.[3] Indeed, there was more at stake than a matter of taste. How one defined the term followed from the attributes one bestowed upon the associates and one's view of the contexts that bring them together. These issues, in turn, informed one's opinion of the prevalence of the phenomenon and its importance for evolutionary change.

Problems in Cost-Benefit Accounting

Some biologists were reluctant to use the term symbiosis if it meant adding mutualism to parasitism. Even those twentieth-century experimentalists who had previously used symbiosis to mean mutualism asserted that this was not a distinct category. When the British microbiologist George Nuttall reviewed the literature on symbiosis in 1923, he found it difficult to imagine that "symbiosis" originated in any other way than through a preliminary stage of parasitism on the part of one of the associated organisms, the conflict between them, in the course of time, ending in mutual adaptation.[4] "Symbiosis" for Nuttall represented a balancing act between two extremes—complete immunity and deadly infective disease. He believed that a condition of perfect symbiosis, or balance, occurred only comparatively rarely because most organisms were either capable of living independently or incapable of resisting the invasion of organisms that were imperfectly adapted to "communal life." He thought it likely that some of the supposed "symbionts" described in the literature would prove to be "parasites" on further investigation.

If benefit had only been assumed for the host, as Nuttall suggested, it was even more difficult obtaining information on what possible benefit was actually gained by the symbionts in many cases. K. F. Meyer raised this issue in 1925 when discussing his experiments on bacterial symbionts in terrestrial prosobranch snails. It was "easy to assume" that the mollusks derive some benefit from the intracellular bacteria as anabolists or catabolists of metabolic waste products. But Meyer found it more difficult to imagine what possible benefit the microorganism derived from the association.[5] He searched the literature for other examples but found only unestablished speculations: "The function of the microscopic 'symbiotes' and their benefit to the host are explained, but little or nothing is said regarding the possible advantage of the microorganisms."[6]

Meyer was not exaggerating. There were microbes (yeast) in the fat bodies of aphids and coccids that were believed to either decompose urates or produce an enzyme that aided in the digestion of sugars; there were microbes in wood-eating insects that were held to aid in the digestion of cellulose; and there were numerous other examples of symbiotes playing a role in digestion. Some investigators suggested that the host secures protein material (or vitamins, according to Portier) by finally digesting the symbiotic algae and fungi. But what was the benefit to the microbe? In the mollusk Cyclostoma the symbiotic bacteria were also eaten by phagocytosis so as to regain some of the substances which the host had previously furnished to the "symbiotes." Nonetheless, some writers, including Paul Buchner, insisted that the bacteria benefited by being protected within the hosts from such drastic influences as heat, cold, and desiccation. Meyer was far from convinced.[7]

Meyer found it difficult to believe that these bacteria should manage to live and propagate better in the cells of animals than outside.[8] The presence of such inhibitory forces would surely be a handicap in the struggle for existence; the microbial "symbiotes" would secure more benefit as true parasites or disease producers. This was perhaps the origin of all so-called symbiotes. Therefore, he argued, the symbiotic communal life was "merely an example of true parasitism or even of disease."[9] This argument, it will be recalled, had been made by Noël Bernard in 1902 to account for the adaptation of orchids to a communal life with fungi; Meyer recommended that it be extended to the "symbiosis" of animals. If this were the case, then such structures as the mycetocytes, bacteriocytes, and mycetosomes in insects would be "a survival of previous profound pathological changes."[10]

Meyer found it easy to imagine how such a communal life in insects might originate from parasitism. He offered the following account. During an early period, the bacteria, fungi, and the like, were probably parasites producing pathologic conditions and disease. Acquired immunity later became inherited, and the microorganisms were gradually gotten under control. The insects do not rid themselves of the invaders because the microbes were transmitted from generation to generation. Still later, the invaders lost all of their harmful effects, and since they secrete enzymes that prove serviceable to the hosts, the conflicts ends in mutual adaptation. In short, the symbiosis between the host and the microorganisms is a phenomenon of parasitism, infection, or disease, which has finally become essential to the existence of the animal.[11]

The view that symbionts originated as parasites was supported by the protozoologist L. R. Cleveland as well.[12] Like the others, he asserted that very few cases of mutualistic symbiosis had actually been demonstrated as such. "The intracellular and intestinal yeast-like and bacteria-like bodies present in all individuals of many groups of insects, and transmitted in most cases from generation to generation through the egg, may or may not be symbiotes," he argued, "—nobody knows." The only crucial method available for determining such a symbiosis was first to break the association, study each partner separately, determine what it can or cannot do when alone, then reestablish the partnership and study the united partners.[13] Although such experimental studies of mutualism were few and far between, Cleveland asserted that

> ideal partnership or symbiosis, that condition in which each party concerned gives as much as it receives, is rarely, if ever, realized; the partnership is usually one-sided, that is one partner does most of the giving. Such associations from the standpoint of the partner that does most of the giving, are near true parasitism; from the standpoint of the partner that does most of the receiving, they are near commensalism.[14]

Not all biologists agreed that symbiosis was due to parasites gotten under control. Ivan Wallin insisted that there was no evidence to indicate that parasitism was the usual manner of development of symbiotic complexes. For Wallin this kind of reasoning was teleological; it invoked intentionality and purposefulness.[15] Parasitism was not a cause and mutualism an effect. In his view, both mutualism and parasitism were effects of relationships. Microbes were no more trying to steal and take advantage of their hosts than they were trying to cooperate. He insisted that in many instances the physicochemical properties of the symbionts entering into a relationship are of such a nature that the terms "parasitism," "infection," and "disease" could not be applied to the relationship.[16]

As Wallin saw it, one had to understand the evolution of symbiosis in the same way as one understood the development of any organism. Symbiosis and parasitism were closely akin, not because the former was derived from the latter, but because both were "end responses in the expression of one and the same biological principle which accounts for the aggregation of cells in complex plants and animals as well as aggregations of organisms."[17]

Hostile Symbiosis

Whether one should understand symbiotic associations in the same terms as one understood the relations among cells of an individual plant or animal remained central to discussions of symbiosis. Some took a genetic view, arguing that "the progressive evolution" of such intimate cooperative relations was primarily restricted to organisms of the same or similar germ plasm. Others maintained that such cooperation was due to the physiological characteristics of organisms and their environmental context. "Commonsense" views of human social relations—of cooperation, exploitation, partners, enemies, masters and slaves, marital relations, and international affairs—continued to be used to explain relations between microbes and their hosts, as well as the tissues of our own individual bodies. The discussion of symbiosis by H. G. Wells, Julian Huxley, and G. P. Wells in their popular synthetic work, *The Science of Life* (1930), is exemplary.

In a chapter entitled "Some Special Aspects of Life," the authors described how the "struggle for existence" among individuals could make and break cooperative relationships, and "how difficult it may be to distinguish service from slavery."[18] At first glance, they argued, it looked as if there were many cases of mutual arrangements by which both partners gain. They mentioned the formation of lichens out of the partnership of alga and fungus, which enabled plants to push farther into the barren places of the earth than they could ever have done otherwise; and how partner bacteria of legumes made atmospheric nitrogen available for their use. They pointed out that a great many animals were green, owing to the presence of single-celled green algae within their tissues. Among these were many of the beautiful radiolarians that float in the sea, the green hydra, and the great majority of reef-forming corals. The animal, they asserted, profits by utilizing some of the carbohydrates built up by the alga's chlorophyll; and the alga profits by utilizing some of the nitrogen which the animal is continuously giving off in its excretions.

They mentioned that all the wood-eating animals and those whose diet is very rich in cellulose were dependent upon symbiotic partners to break down cellulose and wood into digestible substances. Many beetles and beetle larvae, termites, ants, wood-wasps, and gall midges, ruminant and rodent mammals, and certain birds all fell into this category. They pointed out that partner bacteria may be used to generate phosphorescence. Most of the luminous cuttlefish, all the brilliant salps and pyrosomes among the sea squirts, and a few fishes cultivate luminous bacteria in special pockets evolved for the purpose.[19]

There were still other partnerships that were less essential and less intimate, though no less interesting. The authors pointed to the symbiosis between a common species of hermit crab and a sea anemone. The mollusk shell inhabited by the crab is always covered by a pink sea anemone of the genus *Adamsia*. When the crab outgrows its house, it carefully detaches its partner and fixes it onto the new shell. The hermit crab thus obtains a defense force in the shape of the anemone's white stinging threads, or acontia, "covered with batteries of poisonous nettle cells, which it shoots out when alarmed through little port-holes provided for the purpose in its flanks." The anemone, in return, "gains transport and crumbs from the crab's table," as the crab jaws up a piece of carrion. Other crabs live with "evil-tasting" sponges; or they carry sea anemones in their claws "as living knuckle dusters."[20]

From such examples one might imagine that (mutualistic) symbiosis and parasitism were clear-cut biological categories. But Wells, Huxley, and Wells argued that when ecological context and temporal dimensions were added, one could see that such mutualistic behavior was actually driven by hostility. They took the common scotch heather as an example. It was one of the numerous plants that entered into partnership with a fungus and, at first inspection, it appeared to be a perfect case of mutual benefit: the fungus receives carbohydrates made by "the chlorophyll machinery" of the heather, and the heather appropriates some of the nitrogen made available by the specialized chemical activity of the fungus. The heather has become entirely dependent upon the fungus and even needs it as a formative stimulus to its normal development. However, they emphasized, the partnership worked this smoothly and with mutual benefit only in certain environmental circumstances; that is, in the presence of little or no nitrogen salts. If the seeds are grown in the presence of more nitrogen, the fungus grows too vigorously and becomes a parasite instead of a partner, killing the heather seedling. "The regime of mutual help gives place to a regime of exploitation."[21] Such hostility, they maintained, underlay all mutualisms, whether between closely related individuals or different species:

> The phrase "hostile symbiosis" has been used to describe the state of our own tissues—all of the same percentage, all thriving best when working for the common good, and yet each ready to take advantage of the rest, should opportunity offer. There is a profound truth embodied in the phrase. Every symbiosis is, in its degree, underlain with hostility, and only by proper regulation and often elaborate adjustment can the state of mutual benefit be maintained. Even in human affairs, the partnerships for mutual benefit are not so easily kept up, in spite of men being endowed with intelligence and so being able to grasp the meaning of such a relation. But in lower organisms, there is no such comprehension to help keep the relationship going. Mutual partnerships are adaptations as blindly entered into and as unconsciously brought about as any others. They work by virtue of complicated physical and chemical adjustments between the two partners and between the whole partnership and its environment; alter the adjustment, and the partnership may dissolve, as blindly and automatically as it was entered into.[22]

Wells, Huxley, and Wells were careful to point out that our "simple human categories of exploitation and mutual benefit, although useful, are artificial and break down when confronted with the complex and inhuman, or at least non-human, realities of other life."[23] Yet their revelations of "hostile symbiosis" were so twisted and coiled around human social relations that it was difficult to untangle them. Their stories of symbiosis were also about international relations on the eve of National Socialism in Germany, about gender relations—about conflict, instability, and dependence. In their view, the loss of independence of any of the associates was a biologically "degenerate" condition which resulted from a one-sided exploitation. As illustrated in the relations between lower organisms, all partners, whether nations or men and women, had to be wary of exploitation. All relations of any intimacy were "supported, as it were, on a knife-edge. They may so readily over-balance and change into something different and even opposite."[24]

Thus Wells, Huxley, and Wells fit symbiotic relations within both an ecological and human social context. There was no discussion of the role of symbiosis as a general mechanism for constructing new organisms. Though they considered lichens, lupines, and heather to be "compound organisms" and emphasized that many insects relied on their partner microbes, such cases remained exceptions. Ecologists in the United States—even those who emphasized cooperation—were much more critical of the idea that symbiosis could lead to the construction of new compound organisms.

Conflict in Cooperative Ecology

That such classic cases of symbiosis represented compound organisms and mutually beneficial relationships was denied or downplayed by leading ecologists, who emphasized evolution toward a cooperative nature. The views of Warder C. Allee, Alfred E. Emerson, and others at the University of Chicago, are representative. During the 1930s and 1940s their work had been directed toward showing cooperation in intraspecies populations and the integration of the interspecies system, both behaving as superorganismic untis nd subjected to evolution by natural selection.

Like mutual aid theorists before them, they asserted that ethical principles had a biological foundation, that the principle of unconscious mutual cooperation should rank as comparable to the better recognized Darwinian principle of the struggle for existence. There was an evolutionary progress from conflict toward cooperation and toleration. Upholding the concept of the cell state, they maintained that just as the cell in the body functions for the benefit of the whole organism, so does the individual become subordinate to the population.

However, as Donald Worster and Gregg Mitman emphasized, the superorganismic model fell more and more into disrepute in the years surrounding World War II, when critics suggested that it rang of Nazi-like repressiveness, of the subordination of the individual to group conformity, and the possibility was raised of a totalitarian police state based on the same appeals for self-sacrifice to the whole invoked by the organicists.[25] In response to critics, the Chicago ecologists sounded a note of caution against over-integration. They emphasized that their organicist model was intended to be less centralized, less dominated by a single directive force. They sought a compromise between individual freedom and state control: competition was beneficial if it worked for the benefit of the group. Healthy competition between nations, races, denominations, classes, and institutions would encourage progressive social evolution.

In his article "Where Angels Fear to Tread" (1943, published in *Science* as well as five other journals), Allee predicted that when the war was over, all those scientists who helped to win it would "be told in no uncertain terms, as we have been in the past, that war itself was all our doing" and that there would be insistent calls for a moratorium on scientific research lest bigger and more destructive wars be fought.[26] In his view, there would be some truth in the accusations of such "antiscientists." Just as Kropotkin and Reinheimer had argued during World War I, Allee emphasized that "science is not wholly free from war guilt." This was not only because scientists were inventors of tools of mass destruction but because they "have been responsible for giving interpretations to some aspects of Darwinian theories of evolution that provide a convenient, plausible explanation and justification for all the aggressive, selfish behavior of which man is capable."[27] Opposing the views of T.H. Huxley and Spencer, who distinguished human morality from a nonmoral nature, Allee insisted that although one could not deny the existence of strong competitive egoist drives among animals, they had to be kept in "their true place, somewhat subservient to the even more fundamental cooperative, altruistic forces of human nature."[28]

Allee urged biologists to express their natural aggression and competitive urges in terms of "their original function of driving man in his struggle against his enemies among other species of living things."[29] He maintained that scientists of all countries and races "can compete against others for racial, national or personal preeminence in this common task. Competitive drives for worthy ends have real strength."[30] The closer the biological approach was to preventive and curative medicine, the greater the immediate applicability to the war effort. Yet those who studied social evolution, as Allee did, could also be useful:

> Our tasks as biologists, and as citizens of a civilized country, is a practical engineering job. We need to help arrange so that the existing trend toward a workable world organization will be guided along practical lines which accord with sound biological theory. And we must remember always that in such matters the idealist with the long-range view is frequently the true realist.[31]

The work and theorizing of the ecologists at Chicago culminated with the appearance of the great text of Allee, Emerson, Park, Park, and Schmidt, *Principles of Animal Ecology* (1949). They confronted the fascist interpretation squarely and insisted that in the progressive evolution of both human and nonhuman population units, the benefit to the whole system was dependent on the benefits to individuals comprising the group. The surviving system was not one in which the group exploits the individuals or in which individuals exploit the group. Rather, it may be expected that surviving populations will be coordinated under the formula "one for all and all for one." "Lest the concept of the evolution of the biological supraorganism be used to advocate totalitarianism, either of the fascist or communist type," they argued that the evolution of the social unit relied on variation (creative arts and sciences, for example) with social selection acting through optimal competition. "Elimination of nonconformists," they cautioned, "may destroy the degree of social variability upon which progressive social evolution depends."[32]

In the introduction of the text Allee et al. considered the relative neglect of symbiosis to be the result of extreme individualism. It was considered to be more "as an oddity in an egocentric world than as an indication of any general underlying biological principle."[33] Here, one might have thought that symbiosis as a mechanism of evolutionary novelty, a way of synthesizing new compound organisms, would be at least considered, if not highlighted.

Symbiosis was discussed in two chapters of *Principles of Animal Ecology*. In "Biotic Factors in Relation to Individuals," Allee and Schmidt examined commensalism, mutualism, and parasitism.[34] They recommended the term symbiosis in its broad sense to include all aspects of physiological and ecological species partnership, and they used the term mutualism to refer to reciprocal beneficial relationships. Despite philosophers' claims about the value-free nature of science, they recognized that when making such distinctions, one was dealing with short-term operational values.[35]

Allee and Park saw mutualisms in the relationship between plant and animal kingdoms, as well as in the relationship between humans and domesticated plants that could no longer be found in the wild, such as wheat or Indian corn. They believed the lichen to be an example of mutualism, and mycorrhiza was "apparently mutualistic." But they doubted that algae living in protozoa, coelenterates, and mollusks and algae living in association with coral polyps were mutualistic, and they called for their experimental reexamination.

In regard to the importance of such associations in evolution, they referred to the chapter on "Evolution of Interspecies Integration and Ecosystem" written by Emerson. He discussed mutualistic symbiosis in terms of "a tendency of organisms to evolve toward balanced equilibrium with a toleration of other species within the community."[36] In his view, natural selection acted upon the whole interspecies system, gradually resulting in adaptive integration and balance. Thus the interspecies system had evolved the characteristics of what he considered to be an organism, "an ecological supraorganism."[37] Nonetheless, Emerson was more than reluctant to accept any suggestion that any of the classic cases of symbiosis indicated a progressive evolution toward the integration of new organisms, as he believed was the case for intraspecies relations.

In his view, the lichen was not a compound or supraorganism, nor was it an example of mutualistic symbiosis: it was "practically" a parasitic relationship, the fungus holding the alga captive. Through the association with the fungus, algae could exist in places that would otherwise be uninhabitable. Nonetheless, Emerson insisted that all the species of the algae could live independently of the fungus association. He denied the existence of evidence for any modification of the germinal constitution of the alga as a result of the association. Thus he asserted, "lichens cannot be used as a convincing example of the evolution of mutualism."[38]

Similarly, Emerson doubted that both bacteria and legumes had a modified germinal system produced by natural selection as a result of the benefits of their association.[39] There seemed to be an evolutionary adaptation on the part of the bacteria, but there was no reason to believe that the nodules in which the bacteria live represented an adaptive response on the part of the plant; they might be simply caused by agents in the bacteria. Emerson then went further and suggested that the legumes might not be benefited any more than any other plant living in soil containing such nitrogen-fixing bacteria. The relationships of algae living in protozoans, sponges, and coelenterates were also not to be regarded as mutualistic and functional wholes since the animal, in these instances, is not dependent upon the algae. In the case of *C. roscoffensis,* the animal was dependent upon the green flagellates, but in conflict with the views of Keeble, Meyer, and others, Emerson maintained that the reverse was not true: "The flagellates . . . do not seem to have evolved toward dependence upon *Convoluta.*"[40] He conceded that mycorrhizae might be examples of mutualism but suggested that "further experiments were necessary to prove the point conclusively."[41]

As Emerson saw it, symbiosis between insects and protists, which allowed insects to adapt to new surroundings, was a primitive and inhibiting force in the evolution of "higher social functions." Such social progress, as he and his colleagues at Chicago conceived it, resulted from intraspecific solidarity and social adjustment whereby cooperation helped the group achieve greater control over its immediate surroundings, rather than adapting to diverse habitats. To support this view, Emerson pointed to flagellate-harboring termites as studied by Cleveland. In this case at least, Emerson was "forced to the conclusion that the mutualistic relationship has resulted in the evolution of something closely approaching an interspecies supraorganism about as well integrated as parts of an individual organism with selection operating on the system as a functional whole, and favoring greater living efficiency for the mutualistic partners."[42] In fact, he considered the intestinal flagellates of termites and roaches to be "the most remarkable example of evolutionary mutualism known."[43] But this mutualism had been lost in the evolution of "the most advanced termites," the Termitidae.

These "higher termites" had diverged into the largest number of species (1333 described by 1947). Emerson asserted that they were the most successful termites in tropical regions and they had "advanced far beyond their primitive relatives in the integration of their social systems and in their remarkable nest building instincts."[44] Thus he suggested that far from being an example of progressive evolution symbiosis, the flagellate–termite mutualism may have been an inhibiting factor in the evolution of "higher social functions," and once having gotten rid of it, "a great advance and further adaptive radiation of the social system could take place."[45]

In Emerson's view, the evolution toward toleration of other species and adapting toward diverse habitats was secondary to the progressive evolution of higher social functions within species and controlling the "intrasocial environment." He reasoned that progressive evolution was speeded up when it operated through a single germ plasm rather than through two or more germ plasms involved in a symbiotic association. This could help account for "the generalization that the integration of the individual organism and the intraspecies population is far more advanced than the integration of interspecies systems in the community."[46] The evolution of ants, which he believed to have had no interspecies mutualism to start with, had evolved "extreme intraspecies population integration." Mutualistic associations with other species occurred later. Once the more advanced ant and termite had established a degree of social control, he argued, the stage was set for the evolution of mutualistic relations with many other organisms.[47]

Principles of Animal Ecology went through seven reprintings until 1967, at which time, Mitman argues, it had become overshadowed by the ecosystem concept.[48] There was a shift away from an emphasis on studies of the social evolution of intraspecies populations. As the basic functional unit in ecology, the ecosystem concept emphasized the interrelatedness of organisms and their nonliving environment. Any biotic community and its abiotic environment could be considered to be a single functional system of material exchange of energy and chemical substances—a pond, a lake, a forest could provide a convenient unit of study as an ecosystem.

Eugene Odum's *Fundamentals of Ecology,* first published in 1953, became the manual of ecosystem thinking. Odum devoted central chapters to organization at the species, community, and interspecies levels, in which symbiosis and cooperation were discussed. When species of interacting populations had beneficial effects on one another, he argued, it was better to use the word "protocooperation" to distinguish it from the "conscious or intelligent reasoning that is the basis of cooperation between individual human beings."[49] In Odum's ecology, cooperation and competition between species were balanced. The stability of ecosystems relied on the balance. Nonetheless, he continued to lament the lack of studies of such cooperation: "The widespread acceptance of Darwin's idea of 'survival of the fittest' as an important means of bringing about natural selection has directed attention to the competitive aspects of nature. As a result, the importance of cooperation between species in nature has perhaps been underestimated."[50]

Following Darwin's own reasoning about the struggle for existence, Odum argued that mutualisms would be more likely to occur between species and more distantly related kinds of organisms with widely different requirements. Organisms with similar requirements, he argued, were more likely to get involved in "negative interations." He briefly mentioned the classic cases of symbiosis—lichens, fungi and forest trees, bacteria in the root nodules of legumes—each of which he saw as an example of mutualism.[51]

Evolution in the Study of Symbiosis?

Mutualistic symbiosis received brief attention from ecologists and evolutionists of the 1950s and 1960s. If there was "progress" in evolution, there was little in studies of the evolution of symbiosis. In 1952, an English translation of Maurice Caullery's book, *Parasitism and Symbiosis,* appeared. This was the first general account of symbiosis from the viewpoint of an evolutionist, and it remained the only general book on symbiosis for many years. Though the French version had been written thirty years earlier, it was hardly updated, and the main arguments were not modified in the least. They had been directed against the interpretations in Pierre-Joseph van Beneden's *Animal Para-*

sites and Messmates of 1876, in which parsitism, mutualism, and commensalism were constructed in the tradition of natural theology. Caullery's aims were to rid accounts of teleology and anthropomorphism; to show that all cases of mutualistic symbiosis could be accounted for in terms of individual life struggle, domination, and control; and to show that no natural distinction could be made between parasitism, mutualism, and commensalism. As he wrote in the preface: "Commensalism, parasitism and symbiosis are man-made categories which in nature are not discontinuous but are really different aspects of the same general laws."[52]

"Symbiosis," for Caullery, designated "the intimate and constant association of two organisms with mutual relationships assuring them of reciprocal benefits"—a definition he, like so many others, mistakenly attributed to Anton de Bary.[53] Even if one lives at the expense of the other and could be considered a parasite, its metabolism provides the partner with more or less essential elements. Symbiosis involved a physiological relationship and came in two forms. "Ectosymbiosis" involved the regular association of two definite species without the fusion of individuals with one another. In "endosymbiosis"—the typical form of symbiosis—"the two associated organisms interpenetrate and tend to constitute a more or less perfect morphological and physiological unit."[54]

Thus understood, Caullery wanted to show that symbiosis "is not always purely mutualistic and that one of the two associated organisms is in reality more or less parasitic on the other."[55] The most important aspect of symbiosis, in his view, was not mutualism, but the degree of intimacy between the associates. To understand this, he pointed to societies of insects—bees, ants, and termites—in which the complementary functions of the individuals gave the group as a whole a definite collective individuality while at the same time modifying the structure and fate of each individual. Among the many striking features of such organized social structures, he argued, "is the division of labour amongst the associates, localizing certain functions in definite categories and stamping them by definite anatomical and physiological characters, so that castes are formed, forming a unity related to symbiosis itself."[56]

To reveal "illusory anthropomorphisms" in the analysis of ectosymbiosis, he began with accounts of the so-called mushroom gardens of certain ants and termites. The American leaf-cutting ants shred leaves into tiny fragments which they pile up, and on these heaps there regularly develops a mycelium whose hyphae are used as food by these ants. Thus the ants habitually establish mushroom gardens. These associations had been studied in detail since the 1890s by eminent naturalists who observed that when swarming, the ants carried bundles of mycelium in pockets in the hypopharynx and that when new colonies were first being formed, the queen cultivated the fungus; later, the workers assumed the task of caring for the garden. Similarly, the termitarium of a certain family of termites found in French West Africa regularly contains a mushroom garden. In this case, fungi develop on "combs" constructed by the termites with small balls of excreta. Naturalists around the turn of the century who studied this relationship had generally agreed that in the termitarium the fungi were an essential larval food, and that they were indeed cultivated by the termites. (The Chicago ecologists continued to maintain this view and interpreted it as a case of mutualism.)[57] According to Caullery, however, these claims were erroneous and had been refuted during the 1930s.

His colleague at the Sorbonne Pierre-Paul Grassé had claimed that the fungi were not the basic foods of these termites and the existence of mushroom gardens was not a case of intentional cultivation. According to him, the fungal spores were introduced passively with the plant debris out of which the combs were constructed; the combs were nurseries where the termites deposit their eggs. The conditions of temperature and humidity just happened to be optimal for the development of the fungus. Grassé also claimed that the fungi did not constitute a basic food but had at most an accessory value. Thus Caullery insisted that "anthropomorphic tendencies have greatly exaggerated the precision and purpose of the relations between termites and fungi."[58]

To reveal anthropomorphisms in accounts of endosymbiosis, Caullery turned to conflicting interpretations of the lichen. Some believed the fungi were parasitic on the algae: the fungus was the master, the green alga was the slave. Others argued the converse—that the alga was parasitic on the fungus; that is, the algae dominated the fungus. Balanced between these two conceptions stood the idea of mutualistic symbiosis, or a consortium. "On ridding ourselves of verbal phantoms," Caullery remarked, "we find that the question is really one of analyzing, by precise experiments, the relations of the alga and fungus, and of careful comparison of their behaviour in an isolated state and in association."[59]

Noël Bernard's experimental studies of orchids represented the paradigmatic case for understanding lichens and any other case of symbiosis, in Caullery's view. All cases of symbiosis emerged from parasitism. Thus he embraced the conclusion of Alexandr Elenkin, who in 1906 had asserted that the mutualistic conception of lichens would have to be replaced by that of an unstable state of equilibrium: the two associated organisms react differently to the conditions in the external environment and to their variations. Changes in the environment would not be equally favorable to both; and according to the case in question, one or the other would dominate. For the complex to persist, these variations must remain within the limits in which neither of the two organisms succumbs. This conception, Caullery remarked, was equivalent to that of Noël Bernard, "La symbiose est à la frontière de la maladie."[60] All symbiotic associations, even "hereditary symbiosis," could be understood in terms of conflict "terminated by domination of one of the organisms over the other and by a stable equilibrium corresponding to a novel function."[61]

The Organism as Functional Field

If symbiosis meant mutualism, and if mutualism was a "man-made category," an ideal type that did not exist in nature, as Caullery suggested, or at least could not easily be distinguished from parasitism, then one could seriously question whether it was at all useful to continue to use the term in biology. This issue worked itself to the center of discussions of symbiosis during the middle of the century. In 1951, a year before Caullery's book appeared in English, a heterogeneous group of British biologists, among them bacteriologists, pathologists, soil microbiologists, and plant physiologists, assembled to discuss the propriety of the term and its connotations. The soil microbiologist H. G. Thornton framed the problem:

> The term symbiosis has been used with different meanings, and the question of its correct meaning and even of the desirability of its use at all has been debated. The term, indeed, raises the question as to how far it is possible to distinguish a definitely beneficial association between two or more organisms from certain states of parasitism on the one hand and from complex ecological associations on the other.[62]

Though symbiosis could not be defined strictly in terms of mutualism, it was clear to participants that those stable intimate relationships between two or more species could not be understood in terms of strict parasitism either. Some equilibrium was necessary, some exchange of function bonded such symbiotic organisms. F. G. Gregory, a plant physiologist at the Imperial College, London, offered a solution similar to that intimated by Caullery himself: the meaning of the term lay in functional integration and unity.[63] In direct conflict with the views of Emerson, Gregory asserted that the unity of symbiotic complexes was comparable to that of the individual organism. The truth of the claim that any single plant or animal was a unity, he reasoned, was expressed and recognized in the persistence of form within the species. Yet within this unity there was a division of labor between the various organs; each organ by its own activity provides factors essential for the activity of other organs. Thus he suggested that an organism could be regarded as a "functional unity" or "functional field."

The only significant difference between the "unitary organism" and symbiotic associations was that in compound organisms the unity was not dependent upon a single genetic mechanism and the transmission of characters was not secured by the intervention of a single germ cell provided with a complete set of genes. Arguing against those biologists who with Emerson would maintain that this unity was not real because the symbionts could live in isolation under appropriate conditions, Gregory noted that this was also true for the organs of a single "higher plant." For example, in appropriate conditions the root tissues could be grown indefinitely in isolation from the shoot. Grafted plants were comprised of tissues and organs of diverse genetic constitution, and they persisted and functioned as a unit.

Gregory argued that the occurrence of mutual benefits among the associates was not necessary for such unity. "Interlocking of function" and the establishment of an equilibrium could exist in gradations of relationships from mutual advantage to complete exploitation of one organism by the other. Thus he concluded that there was not much to be gained by postulating a mutual relationship. "The value of the concept of symbiosis resides in the widening of the concept of organism to include heterogeneous systems overriding the limitations of genetic uniformity, and the supplementation of the concept of a structural unity by that of a 'functional unity' or functional field."[64]

Other participants at the meeting agreed that there was not much to be gained from using the term symbiosis in the sense of mutualism. The extent to which symbionts benefit from their association could in each case depend upon the overall environmental context. Symbiosis therefore represented "a steady state of dynamic equilibrium or stability maximum established between, rather than a rigid biological obligation imposed upon, the symbionts." This point was obscured when such terms as symbiosis, mutualism, commensalism, and the rest, were employed "as though they stood for inviolable legal agreements contracted between acquiescent or hostile parties."[65] There were also grave practical difficulties of assigning some complex associations to one of the categories when many different kinds of symbionts were involved. Any attempt to deal with such complex interactions "as a simple problem of accountancy based on profit and loss encountered insuperable obstacles."[66] "In view of these difficulties" one commentator remarked,

> it may well be necessary, as Professor Gregory has recommended, to discontinue the use of the word symbiosis, substituting for it the more appropriate term "functional field." . . . If this were done, questions could profitably be raised regarding the degree of integration of symbiotic associations considered as a function of the intensity of the field established and of the internal and external resistances surmounted in its establishment.[67]

Bacteriologists at the meeting also insisted that no useful purpose was served by a distinction between symbiosis and parasitism, that cost-benefit analyses of symbiotic associations were bankrupt. Some claimed that the fundamental association in symbiosis was between two cells, not between bacteria, and for instance, a plant or animal. In these cases there was no reason to suppose that the host cell derived benefit, and biologists were not yet in a position to judge such matters. As the Oxford bacteriologist Sir Paul Fildes quipped, "The fundamental association, whether thought to be beneficial to the cells or not, is chiefly important because it provides a subject for biochemical study which might lead to secondary effects beneficial for us."[68]

Thus participants at the meeting concluded that there were both logical and practical reasons for broadening the definition of symbiosis to embrace parasitism or, to put it another way, to avoid such concepts, and labels based on cost-benefit analysis altogether. This was hardly the end of attempts to define symbiosis. Discussions of definitions opened almost every paper and volume on symbiosis during the 1960s. Images from human social relations continued to be employed. As one writer remarked in 1966: "Symbiosis is like marriage—a mutual exploitation." This epigrammatic definition describes perfectly many symbiotic relationships.[69]

One Germ Plasm—One Organism

Despite Gregory's recommendations for widening the concept of organism to override the limitations of genetic uniformity, niether he nor any of the other participants at his meeting discussed the idea that all plants and animals are symbiotic complexes. In all of the discussions about the meanings and mechanisms of symbiosis, the idea that chloroplasts and mitochondria were symbionts was either ignored or dismissed. Symbiosis remained an exceptional phenomenon. This was true for biologists of every disciplinary background and of every major evolutionary persuasion, whether Darwinian or non-Darwinian. Maurice Caullery, like many neo-Lamarckian biologists and many embryologists, maintained that Mendelian genetics was concerned with only trivial differences that did not extend beyond the species. Nonetheless, he transcended such evolutionary debates when he asserted that such cases of symbiosis as lichens, mycorrhizas, bacteroids in nodules on roots, yeast in insects, and luminous bacteria were "deviations from the normal."[70] Portier's failed attempt to demonstrate that mitochondria are symbiotic bacteria, in Caullery's view, did not "dispel all possibility of reality for ideas for cellular dualism and the existence of autonomous organisms." Nonetheless, he believed that it was "just as natural, if not more so, to attribute to the cell itself the faculty of accomplishing the essential functions of life."[71]

In a chapter on the "A B C of Genetics," Wells, Huxley, and Wells mentioned symbiosis when considering the idea that the cytoplasm may be concerned with "species inheritance" while the nucleus may be concerned only with "variety inheritance" and when discussing evidence for the non-Mendelian inheritance of chloroplasts in some plants. In regard to the latter, they wrote: "It has even been suggested that these plastids are organisms, independent lives within the cells, just as the algae in lichen are captive lives within a larger organism; but nobody has yet isolated a plastid and seen it grow apart from the containing cell."[72]

As for the dualism of cytoplasm and nucleus, they recognized that it was possible to regard the chromosomal mechanism as a later evolution, coexisting side by side with a more ancient mode in plants and animals. However, they dismissed the terms "species" and "variety" inheritance as being unjustified. And their statement about the origin of nucleated cells left no room for symbiosis and little room for cytoplasmic inheritance:

> Presumably when the first bacteria evolved upwards into creatures with nuclei, an elaboration that we must stress as the greatest stride ever made in the evolutionary march, they did so by a process of concentration—by collecting all this scattered stuff [chromatin] together into a single central blob. If this process was at all complete, no hereditary material would be left in the cytoplasm outside the nucleus.[73]

Without mentioning mitochondria or any of the established case of hereditary symbiosis, they dismissed "non-Mendelian heredity, that is to say, heredity through any other agency than the [nuclear] genes, as a process trivial in extent and at present too obscure for practical consideration."[74]

This nucleocentric view of heredity, so prevalent among English-speaking geneticists, was challenged during the decade following World War II when new techniques, aims, and concepts came to revolutionize much of genetics. By the middle of the century, and rather unexpectedly, the same discussions about broadening the concept of the organism to include symbiosis came to the forefront of genetics.

10

Organisms and the Edge of Disciplines

This review is dedicated to the reconciliation of the attitudes that plasmids are symbiotic organisms, and that they comprise part of the genetic determination of the organic whole. The conflict may arise in part from fixed conceptions of the scope of the organism.

JOSHUA LEDERBERG, 1952[1]

Concepts of hereditary symbiosis rose to the center of genetic discourse during the decade following World War II when new organisms—protozoa, algae, fungi, yeast, bacteria, and their viurses—were domesticated for genetic use and when geneticists began to focus on the nature of the gene and how genes control biochemical processes, as well as the problem of cellular differentiation, which for half a century had been largely excluded from genetics. A new generation—combined with the immigration of chemists and physicists into the domain—transformed the practice of genetics.[2] New countries entered the field while others abandoned it. In France, small but very influential groups emerged at the Pasteur Institute and the Institute of Physicochemical Biology, Paris, working on problems of gene action and regulation. In the Soviet Union, Lysenkoism arose, repudiating the principles of genetics in a brutal wholesale way.[3] In the West, postwar technological innovations in genetics were coupled with somewhat revolutionary concepts which threatened previously guarded genetic doctrines, not just about the nature of the gene, but also of heredity *sensu lato,* and the definition of the organism itself. The Mendelian chromosome theory was extended to include microorganisms. At the same time, new genetic evidence for cytoplasmic heredity emerged.[4]

Several leading biologists, most prominent among them Tracy Sonneborn and Joshua Lederberg in the United States, Boris Ephrussi, André Lwoff, and Philippe L'Héritier in France, Jean Brachet in Belgium, and C. D. Darlington in England, challenged the "nuclear monopoly" of the cell. Genetic demonstrations of non-Mendelian inheritance were linked with various cytological studies of cytoplasmic bodies long neglected by Mendelian geneticists. They included centrioles, which in most animal cells (and exceptionally in plant cells) could be seen at the poles of the mitotic spindles; kinetosomes, which could be found at the base of cilia and flagella of many unicellular organisms and of ciliated and flagellated cells of metazoa; mitochondria, ubiquitous elements present in any living plant or animal cell; and plastids of several sorts—chloroplasts and leucoplasts—which assumed various functions.

While some geneticists associated the new evidence for cytoplasmic heredity with problems of development and cell differentiation, as the long-awaited link between genetics and embryology, others continued to trivialize it. The conflict over the relative importance of the nucleus and cytoplasm became heated. The idea that the cytoplasm could be in control of the fundamental characteristics of the organism was rehabilitated, and the evidence for extranuclear inheritance became entangled in the Cold War dispute surrounding the Lysenko affair.[5] It was in this context, of debates over the scope and significance of cytoplasmic heredity, that geneticists discussed the meaning and scope of symbiosis for heredity and evolution. While offering a synthesis between heredity and development, geneticists' studies of cytoplasmic inheritance also brought them closer to infections and symbiosis.

Plasmagenes, Viruses, and Viroids

During the 1940s and 1950s all extranuclear hereditary determinants were discussed under the rubric of "plasmagenes." Plasmagenes were conceived of as self-reproducing genetic entities, varying in size from microscopically visible particles to submicroscopic particles of the same order of size as nuclear genes. Some of the larger elements, such as chloroplasts, were thought to contain plasmagenes within them. The chief theoretical importance of plasmagenes was how they were invoked to account for cellular differentiation. Unlike nuclear genes, they could be sorted out during cell division under the influence of the environment and offered one of several cytoplasmic mechanisms to account for cellular differentiation in the face of nuclear equivalence.

Sonneborn and his colleagues investigated various cases of non-Mendelian inheritance in the ciliated protozoan *Paramecium.* Their work on the "killer trait" provided the exemplar of the interaction of plasmagenes and the environment. Some strains of paramecia produce a poison that kills paramecia of other strains. Sonneborn first reported the the killer trait in 1943 and reasoned it to be due to a cytoplasmic genetic substance he called *kappa.* The effect on *kappa* of environmental conditions was striking. The growth rate of paramecia could be controlled by such environmental elements as nutrition and heat. If cells were kept in a medium where fission was rapid, they tended to multiply more rapidly than the cytoplasmic factor *kappa,* and eventually the large majority of cells ceased to be killers. In this way, the concentration of these cytoplasmic components within cells of a clone, and thus the character of the cell (killer, sensitive, resistant), could be controlled by environmental conditions. The behavior of *kappa* was similar to that of chloroplasts in certain unicellular green algae. For example, when *Euglena mesnili* is cultivated in the dark, the rate of division of its chloroplasts is slowed down relative to the organism divisions, resulting in a decrease in the average number of chloroplasts per organism. Eventually organisms arise which are irreversibly devoid of chloroplasts.

This model seemed to be applicable to a heterogeneous population of mitochondria as well.[6] Some of the first genetic evidence indicating mitochondrial heredity was reported by Boris Ephrussi and his colleagues at the Institute of Physicochemical Biology in Paris. It concerned slow-growing, respiratory-deficient colonies of yeast widely known as "petite [colony] mutations." Ephrussi allied his work to plasmagene theory and in 1953 published a small book, entitled *Nucleo-Cytoplasmic Relations in Microorganisms,* in which he offered various cytoplasmically based models of cellular differentiation and reconsidered the idea that the cytoplasm determined the fundamental characteristics of the organism.[7]

At the Pasteur Institue, André Lwoff brought forward cytological studies of the life history of kinetosomes in ciliates, which he and his mentor Edouard Chatton had carried out in the 1930s. In a small book entitled *Problems of Morphogenesis in Ciliates* (1951), Lwoff developed a theory of the nature and behavior of kinetosomes as plasmagenes with far-reaching implications for differentiation and morphogenesis. Kinetosomes, he argued, were able to multiply independently of the nucleus and gave rise to chains or rows of kinetosomes (kineties) along the cell surface or cortex. The metabolism of the "host," light, and temperature could cause variations in the relative multiplication rates of kinetosomes in a way similar to the behavior of *kappa* and chloroplasts.

Lwoff maintained that kinetosomes had another special property which others in France had previously accorded to mitochondria: cilia, centrioles, and various other fibers and organelles, including blepharoplasts, parabasal bodies, and basal granules, could be organized into different structures and systems according to their position in the cell. For Lwoff studies of the life history of kinetosomes represented a visible confirmation of the theoretical conclusions of embryologists that development is largely effected by the cytoplasmic particles operating under the influence of a morphogenetic field. "The morphogenesis of a ciliate," he declared, "is essentially the multiplication, distribution and organization of populations of kinetosomes and of the organelles which are the result of their activity."[8]

Throughout the decade following World War II, the origin of such plasmagenes and their significance for heredity and evolution remained controversial. Some geneticists believed they might originate as products of nuclear genes which migrate to the cytoplasm and reproduce there, similar to the suggestion of de Vries, Weismann, and other nineteenth-century theorists. Others saw them as semiautonomous, self-duplicating, mutable, cytoplasmic particles which depended on the nucleus for their maintenance and normal functioning but not for their origin or for their specificity.[9] Still others dismissed some of the evidence for plasmagenes as due to parasites or symbionts of extrinsic origin, and denied their generality as being of little significance for heredity.

The debate over the symbiotic interpretation of plasmagenes began in 1946 when the American geneticist Edgar Altenburg published two papers, "The Viroid Theory in Relation to Plasmagenes, Viruses, Cancer and Plastids" and "The Symbiont Theory in Explanation to the Apparent Cytoplasmic Inheritance in Paramecium."[10] The attitudes expressed in these papers reflected well geneticists' views toward the nature of the gene and its similarities to viruses. Altenburg was a former student of H. J. Muller, who that year was awarded a Nobel Prize for his work on gene mutation. In 1922, shortly following the seminal work on bacteriophages by Félix d'Herelle, in a statement that has often been regarded as prophetic, Muller had compared viruses to genes in their ability to multiply and mutate. He did not suggest how they might be phylogenetically related, nor did he want to engage in such a discussion. Instead, he argued solely on practical grounds that perhaps the study of viruses would be useful in revealing the nature of the gene itself.[11]

For the most part, viruses had no history. They were at best technological instruments in the hands of geneticists, perhaps, as Muller suggested, useful for understanding "our genes," for they were "naked genes." During World War II studies of bacterial viruses (bacteriophages) grew in the United States under the German emigré physicist Max Delbrück. Several members of the new school of phage workers were, like Delbrück, trained in the physical sciences, and they regarded the phage neither from a medical nor from a basic bacteriological point of view, but "merely as a tool for investigating hereditary processes."[12]

For hundreds of years the term "virus" had been used to designate the principles or agents of transmissible, contagious, infectious disease. Some microbiologists of the 1950s still employed this meaning. Those agents which were not visible with an ordinary microscope were called filtrable viruses while the visible agents of infectious diseases, the protists or microbes, lost their title of viruses.[13] However, there was no more consensus about the origin of viruses than there is today.[14] Altenburg's first paper of 1946, which interpreted plasmagenes as viruses, represented an attempt to give viruses a history.

Altenburg supported his suggestion that ultramicroscopic plasmagenes were symbiotic viruses with evidence of symbiotic bacteria harbored in the fat bodies of cockroaches and termites. He referred to them as "bacteroids" to distinguish them from disease-causing bacteria: they were "symbionts," which he believed to be necessary for some cell process—perhaps for the building up of some vitamin or for some other metabolic process.[15] If bacteria—that is, bacteroids—could be useful to their hosts, he reasoned, then perhaps viruses (or at least a close relative) could be too. Thus Altenberg postulated the existence of *viroids,* which would be akin to viruses but which would be useful symbionts occurring *universally* within the cells of larger organisms.[16] Ordinarily biologists would be unaware of their existence, but if they mutated, he argued, they would give rise, for example, to the killer factor described by Sonneborn in the cytoplasm of *Paramecium.* Such viroids could also be related to all cancers. Even those cancers not due to infection from the outside might be derived by mutations of a viroid normally present as a symbiont in the cell cytoplasm.

Altenburg offered an account of the origin of such symbiotic viroids from the origin of life to the present. Since they were neither plant nor animal, he made a new separate kingdom for viruses: the Archetista, which had been free-living in an early stage of evolution—the naked gene stage—and were the dominant form in the Archeozoic. With the emergence of bacterialike organisms, some Archetista became symbiotic while others continued as free-living organisms. The small size of the Archetista would have rendered them more adaptable than the bacteria as intracellular symbionts, especially in the earlier stages of evolution, when most cells were not yet much larger than the bacteria themselves. Thus Altenburg reasoned that "the Archetista would have pre-empted the field as intracellular symbionts. Hence their expected universality as symbionts in the larger organisms."

Unlike many others who had offered accounts of the origins of symbiosis, Altenburg did not believe that symbiosis generally evolved from parasitism. He argued against this view on the grounds that affected hosts would be as a rule killed before the parasitic agent could change to a symbiont. Viroids would continue from "amoeba to man" since they would have an indispensable role in the economy of the cell. They would then have served as the "reservoir" for viruses in the course of evolution. Viruses for diseases of comparatively ancient origin (such as yellow fever and smallpox), he dubbed *paleoviruses*. Cancer viruses, on the other hand, would arise anew in each cancer case from a mutated viroid. These *neoviruses* would be more closely related to viroids than an ordinary virus (as for smallpox).[17]

Altenburg's viroid theory was supposed to account for cytoplasmic heredity for which no visible body had been identified. He also acknowledged that chloroplasts in plants were semiautonomous bodies and entertained the possibility that they might have come into the plant cell as symbionts, like the algae in lichens. But he claimed that there was "no evidence for this view," and he favored the opinion of his mentor, Muller, who had suggested that chloroplasts antedate the nucleus and simply have not become incorporated in it. In other words, they arose from chromatin orginally present in the protoplasm of primitive cells before the major portion had become surrounded by a nuclear membrane.[18]

Altenburg was not the first geneticist to associate invisible plasmagenes with viruses. C. D. Darlington had made the suggestion two years earlier.[19] However, in his theory, plasmagenes were well-integrated parts of the gene system, but they could mutate and become viruses and cause cancer as well as infectious diseases. But Altenburg denied the plausibility of this view and dimissed Darlington's concept as being reminiscent of the ancient and discredited theory of spontaneous generation.

In his second paper of 1946 Altenburg offered an alternative to his viroid interpretation of *kappa*. One of Sonneborn's students, John Preer, had reported that *kappa* was visible in appropriately stained strains of *Paramecium aurelia* and was rather limited in number to about 250 per organism. Now that *kappa* particles could be counted, Altenburg gave them a new association. There were about the same number of green symbiotic bodies in *P. bursaria,* but these green bodies were not present in all strains of *P. bursaria.* Perhaps the colorless *kappa* was derived from loss of pigmentation in what were originally green bodies. But whether viruses or not, one issue was clear to Altenburg: "The *kappa* factor cannot represent any important element of the gene system proper of the Paramecium and it is therefore more probably a symbiont."[20]

Many geneticists came to suspect that *kappa* was some kind of symbiont, but its true identity remained a subject of controversy. The symbiont interpretation of *kappa* and perhaps other so-called plasmagenes became more plausible to geneticists when Sonneborn and his colleagues showed that *kappa* could indeed be transmitted artificially by infection under special laboratory conditions. This was also true of the cytoplasmic genoid *sigma* in *Drosophila* investigated by Philippe L'Héritier. But L'Héritier refused to put the genoid in any definite category on the grounds that its classification was "doomed to remain so much a question of definition and personal feelings."[21]

Inherited Symbionts: Bad Pennies

In debates about the relative importance of the nucleus and cytoplasm in heredity, when few cases of cytoplasmic inheritance were demonstrated, the infectious nature of *kappa* and other plasmagenes was a liability. It only lent more credibility to those geneticists who upheld the exclusive or predominant role of the nuclear genes in heredity and who argued that, with few exceptions, one could dismiss the idea that cytoplasmic genes or gene complexes played any normal role in heredity.

In September 1950, the Genetics Society of America held its Golden Jubilee at Ohio State University. H. J. Muller gave an account of the development of the gene theory. Following a discussion of whether or not Mendelian gene differences could account for macroevolution, he summarized the arguments against semiautonomous cytoplasmic bodies playing any essential role in the physiological processes of animal cells. They included (1) the extreme rarity with which illustrations of such inheritance have been found in animal material, in contrast to the thousands of Mendelian differences found in them; (2) the fact that the agents have been proven to be able to pass as infections from one cell to another; and (3) the lack of a fundamental basis for distinguishing between these and cases of undoubtedly parasitic or symbiotic microorganisms or viruses of exogenous derivation. Muller concluded, "These are points which, taken together, would appear to argue for most or all of these agents in animals having at one time arrived as invaders; for their still constituting, in a sense, an adventitious part of the inheritance, and for their tenure usually being insecure, as compared with that of the native chromosomes."[22]

Muller did not believe that such cytoplasmic "invaders" in animals could ever evolve to be normal cellular constituents comparable to the chromosomes. He referred to none of the literature pertaining to "hereditary symbiosis" or to the possibility that normal cytoplasmic components such as choroplasts, mitochondria, and centrosomes might have evolved as symbionts. He simply reiterated his previous suggestion that chloroplasts were "derived from some of the genes which were originally present in the protoplasm of primitive cells before the major portion of the chromatin had become sequestrated within a nuclear membrane."[23]

Muller's criticisms did not go unchallenged. Proponents of cytoplasmic heredity drew upon technical, theoretical, and social components of science to account for the lack of evidence of cytoplasmic inheritance compared to the nucleus.[24] The relative lack of genetic evidence of cytoplasmic inheritance compared to Mendelian inheritance was not a reflection of nature; it was simply a reflection of the relative ease of investigating Mendelian phenomena with the techniques available and of a lack of willingness on the part of most geneticists to investigate and report non-Mendelian phenomena when they detected them. Geneticists would normally dismiss and not report messy cytoplasmically inherited traits when they found them. Instead, they would focus on those cases which were easily publishable. According to this argument, then, the facility of Mendelian genetic techniques reinforced a reward system in science that favored orthodoxy and selected against investigation of exceptions. Theoretical arguments reinforced sociotechnical ones: if cytoplasmic genes controlled fundamental or crucial cell functions, then mutations in such traits would often lead to death of the organisms which possessed them and therefore would seldom be detected.

The common criticism that cytoplasmic bodies lacked a mechanism comparable to the chromosomes for transmitting themselves safely from one generation to the next was equally rebuffed. Even *kappa* particles which were randomly distributed from one cell generation possessed a mechanism that would ensure their continuity. Thus the reproductive rate of *kappa* particles was regulated by their concentration in the cell. When the concentration of *kappa* dropped, the reproductive rate increased; when the concentration of *kappa* rose, the reproduction rate fell. There seemed to be a physiological regulation of some kind. Kinetosomes, however, remained fixed on the cell cortex of ciliated protozoa. There, they duplicated and at fission every organism received its full allotment of kinetosomes. Still other cytoplasmic structures, such as chloroplasts, mitochondria, and centrioles, could be as precisely distributed as were chromosomes. This is not to say, however, that the reproduction of plasmagenes was always synchronized with nuclear and cell division. The rate of reproduction of *kappa* and chloroplasts, for example, could be controlled directly by environmental conditions, such as nutrition, light, and heat.

Then there was the ever-pressing criticism that the cases of cytoplasmic inheritance were due to parasites or symbiotic microorganisms or viruses of exogenous origin and therefore did not, in the words of Muller, "form an essential part of the genetic constitution of animals." The "petite" mutations in yeast studied by Ephrussi and his colleagues did offer genetic evidence that cytoplasmic heredity might exist in animals insofar as they were associated with mitochondria. By 1950, even Sonneborn was willing to admit that it was possible that *kappa* and *sigma* were symbionts. However, this was not to say that they (and any other plasmagenes) were merely infectious parasites or little genetic significance. He pointed out the *kappa* was well integrated into the physiology of the strains of *Paramecium* that possessed them. *Kappa* particles could mutate like genes and diverse mutant *kappas* were maintained and multiplied in cells. They therefore possessed the properties of the building blocks required by the process of evolution which Muller had reserved solely for nuclear genes in animals.

Moreover, both *kappa* and *sigma* were infectious artificially by injection—only under laboratory conditions which could scarcely be imitated in nature. They could not be dismissed simply as parasites or symbionts of no genetic significance. They "are of genetic significance," Sonneborn argued, "because they normally transmitted only by heredity." Sonneborn did not suggest that all self-perpetuating cellular bodies were also once symbionts. Instead, he used this idea to mock those who would dismiss evidence for plasmagenes on this basis:

> One cannot but be impressed by the fact that practically every self-duplicating structure occurring within cells has at one time or another been considered a symbiont or parasite. For example, chloroplasts and mitochondria have been interpreted as symbionts. . . . Even the nuclear genes have not been spared.[25]

To call a particle a symbiont amounted to name calling. Sonneborn's and Muller's arguments represented opposite sides of the same coin: if one was willing to "spare" nuclear genes from this interpretation, one should rightly "spare" plasmagenes as well. The inheritance of symbionts was a bad penny which held no currency in genetics. One's rightful inheritance came in the form of endogenous genes that were transmitted solely sexually.

Changing Concepts of Heredity

Controversies over the interpretation and scope of cytoplasmic heredity, often centering on the diagnosis of virus versus plasmagene, remained at the forefront of many symposia of the late 1940s and early 1950s. Though Sonneborn responded to Muller in Muller's own terms, the idea that all cases of cytoplasmic inheritance were due to symbionts *and* that symbionts had to be considered as part of the genetic constitution of the complex organism was growing among some leading geneticists.

Throughout the first half of the century, geneticists had restricted the concept of heredity to suit their practice. Inheritance as borrowed from jurisprudence was the transmission of rights and properties from parents to offspring. Geneticists transformed the metaphor to the transmission of genes from one generation to the next. According to the evolutionary synthesis of the 1930s and 1940s, evolution resulted from a gradual accumulation of almost imperceptible differences between individuals of a species. Undirected gene mutations and genetic recombination represented the primary sources of evolutionary variation. The individual organism was well defined and protected against invaders. Infection was opposed to inheritance, as disease was to health, as poverty to wealth, discontinuity to continuity. The true germ line was pure, protected from the mundane realities of life and free from the kinds of contaminations that might invade otherwise healthy tissue. An infectious agent was at best a harmless foreigner, carrying out special duties in certain circumstances; most often, it was a thief which wormed its way in to rob one of one's rightful inheritance.

Yet in microbes themselves none of this held up. They could multiply without sexuality, and there was no separate germ line in microbes. In bacteria, moreover, there were other means besides sex for transmitting genetic material from one generation to the next. By the early 1950s it was becoming evident to geneticists that infectious viruses could act as vehicles of inheritance in bacteria. Their study, when combined with studies of cytoplasmic inheritance in other organisms, reinforced the idea that symbiotic microorganisms could become well-integrated into, and form an essential part of the genetic constitution of the organic whole.

The celebrated population geneticist Theodosius Dobzhansky mentioned the possibility that all plasmagenes were symbionts in the third edition of his well-known book *Genetics and the Origin of Species* (1951). By that time, Dobzhansky had begun to entertain the idea that populations had organismic properties and were units of selection, as developed most prominently by Warder Allee and his colleagues at Chicago.[26] When briefly discussing parasitism and mutualism, he argued that disoperative (parasitic) relationships between species were unstable in an evolutionary sense and would "tend to disappear and to be replaced by cooperation and mutualism."[27] He quoted from the Australian virologist MacFarlane Burnet the reasons why a fatal infection is disadvantageous for the survival of the parasite; how rickettsia, old associates of ticks, cause them no harm and are transmitted from one generation to the next through the egg cytoplasm; how the virus *Herpes simplex* has accomplished an "almost perfect" evolutionary adjustment to the human host. "It is not surprising" he wrote, "that the Killer particles were at first described as plasmagenes, and indeed plasmagenes may well be symbiotic organisms which have become integral parts of the cells of the host as a result of a long association."[28] Nonetheless, plasmagenes, whether symbionts or not, were still relatively rare in Dobzhansky's view. As he put it: "True cytoplasmic inheritance, caused by the presence of self-reproducing entities in the cytoplasm, occurs in some plants and microorganisms, as well as in a few animals."[29]

C. D. Darlington made similar remarks. Though he had previously viewed plasmagenes as nuclear products, by 1950 he was "gradually being driven to conclude that there is a wider range of cytoplasmic determinants of greater power than our predecessors had dared to suppose."[30] And just as it had been necessary to adopt new terms for Mendelian or nuclear genetics to avoid confusion with old ideas, one had to adopt new new terms for cytoplasmic determinants. However, one could not do so with the same precision. One could not make a hard and fast distinction between a virus, a symbiont, and a plasmagene. For "what is heredity and variation for a protozoan is development and differentiation for a higher organism. And what is heredity for a particle of Rickettsia transmitted through the egg of a bug is infection when the same particle is transmitted by the bug to man."[31]

Darlington extended his previous theory that plasmagenes could mutate into viruses to include the opposite change from virus to plasmagene. The *kappa* particles of *Paramecium* were indicative of such a possibility. In his view, the difficulty of distinguishing between a plasmagene and a virus in the case of *kappa* was even more severe than the difficulty of distinguishing plants and animals among microbes:

> The fact that some difficulty arises in the distinction is what makes the situation so significant, more so than the difficulty arising in the distinction between plants and animals. It means that the pathologist cannot dismiss a particle as "only a plasmagene" and that the geneticist cannot dismiss a particle as "only a virus." Both have to take both kinds of particles as their own serious concern.[32]

Darlington prophesied that cytoplasmic inheritance would in the future enable geneticists "to see the relations of heredity, development and infection and thus be the means of establishing genetic principles as the central framework of biology."[33]

Lysogeny, Symbiosis, and Technoevolution

The most elaborate attempts to resolve the debate between plasmagene versus virus and to construct a unified concept of heredity in which a cytoplasmic constituent could be both a symbiont and a part of the hereditary constitution of the organism as a whole was made by Joshua Lederberg. Though only in his mid-twenties, Lederberg was a well-known figure in genetics by the early 1950s. In 1946, at Yale University, he and Edward Tatum showed that bacteria had genes and could be cross-bred and exploited for genetic analysis. After receiving his doctorate in 1947, he accepted an appointment at the University of Wisconsin where he and his collaborators continued to explore the ramifications of genetic recombination in bacteria combined with the study of bacteriophages. For his studies of the organization and recombination of genes in bacteria, he was awarded a Nobel Prize with Tatum and George Beadle in 1958.

During the early 1950s, Lederberg coined some of the central terms for the study of bacteria and their viruses, such as "plasmids" and "transduction." These terms, and the study of the bacteriophage *lambda* in *Escherichia coli*, are today perhaps best known for their application to techniques of genetic engineering.

The development of genetic engineering derived from scientists' observations of natural recombination mechanisms in bacteria. This involved "transformations," that is, conjugation and most analogously transduction—the transfer of genetic material from one cell to another by a virus or viruslike particle. Though today this work is usually seen as providing foundations for recombinant DNA technology,[34] during the early 1950s, these concepts and terms were embedded in debates over the scope and significance of cytoplasmic inheritance and the role of symbiosis in heredity and evolution. This is not to say, however, that Lederberg's arguments about the nature of viruses and the significance of "hereditary symbiosis" were driven solely by purely evolutionary concerns. Like Muller's concept of viruses as naked genes thirty years earlier, they were equally clothed in practical arguments about the use of such entities as technologies for studying the nature of the gene and gene action. We might call them "technoevolutionary arguments."

Beginning in the late 1940s, Lederberg associated *kappa* particles in paramecia with two well-known studies of what he considered to be "infective heredity" in bacteria. The first was the pneumococcus "transformation" investigated by the immunologist Oswald T. Avery and his colleagues at the Rockefeller Institute for Medical Research. This work was based on experiments of the British pathologist Frederick Griffith who, in 1928, had reported that nonvirulent strains of *Pneumococcus* could be transformed to the virulent type when mixed with heat-killed virulent strains. Subsequent investigations by Avery and his colleagues culminated with their startling report in 1944 that the active material in the "pneumococcus transformations" was DNA. The noninfectious bacteria picked up loose threads of DNA that had been released from the heat-killed lethal strain. Furthermore, the acquired DNA and its power to infect were transmitted to the progeny of the transformed bacteria. In 1949, Lederberg argued that "from purely mechanical considerations it would seem most likely that the transforming agents are incorporated into a cytoplasmic system like that of *kappa* to perform such functions."[35]

Lederberg saw a further analogy between *kappa* and lysogenic bacteria in which a latent virus or "prophage" is transmitted as a hereditary quality. The inheritance of such a virus would remain undetected except that occasionally it develops into an active virus, kills its host, and is released into the medium. Once freed, the virus can infect other bacteria and behave alternatively as a lethal parasite or render the bacterium lysogenic.[36] In 1951, Esther Lederberg stumbled across the bacteriophage *lambda* in a strain of *E. coli* which Joshua Lederberg had previously worked out for crossbreeding analysis. The Lederbergs referred to it as an "enduring symbiosis" and investigated it "as a model system for the study of cytoplasmic heredity."[37] In fact, they named the lysogenic virus they investigated in *E. coli* K12, *lambda* since, on the analogy of *kappa* particles in *Paramecium,* the symbiotic virus in bacteria might also behave as an extranuclear factor. However, the very possibility of such a symbiosis of virus and bacteria had been denied for many years.

The history of lysogeny, the observation of the phenomenon, and its denial is a well-told mythical tale among bacterial geneticists. The myth begins with the views of Félix d'Herelle, who coined the term "bactériophage" in 1917. D'Herelle had observed what others later claimed to be lysogenic bacteria, but, according to the phage geneticist Gunther Stent, he dismissed the phenomenon as due to contamination of the bacterial culture through improper technical procedures during the purification procedure.[38] Stent has emphasized, "That any virus could live in such symbiotic coexistence with its host was, however, in flagrant contradiction with what contemporary bacteriologists, including d'Herelle, imagined to be the morbid essence of virus reproduction."[39]

However, as discussed in Chapter 7, in the mid-1920s d'Herelle explicitly discussed the perpetuation of mixed cultures of bacteriophages and bacteria, or lysogenic bacteria, in terms of symbiosis—as microlichens. Indeed, the transformations associated with such phage-resistant bacteria, as well as other examples of transformations accompanying symbiosis in plants, encouraged him to believe that symbiosis was largely responsible for evolution. Portier also had interpreted bacteriophages as symbiotes, as had Meyer and Wallin during the 1920s. In Australia in the 1930s, studies of the virologist MacFarlane Burnet and his collaborators led him to postulate that "each lysogenic bacterium carries in intimate symbiosis one or more phage particles which multiply by binary fission *pari passu* with the bacterium."[40] At the Pasteur Institute a third alternative was entertained: bacteriophages were of endogenous origin. In 1925, Eugène Wollman, collaborating part-time with his wife, Elisabeth Wollman, put forward the hypothesis that some genes are transmitted from cell to cell by the external medium, a phenomenon he called "paraheredity."[41] The Wollmans' work was tragically interrupted at the end of 1943 when they were arrested in the Pasteur Institute and taken by the Gestapo to Auschwitz, never to be heard from again.

The views Stent attributed to d'Herelle more accurately reflect those of the Delbrück school in which Stent was trained. Delbrück and his school flatly refused to believe in a viral–bacterial symbiosis, maintaining that a bacterium infected by phage bursts minutes later, releasing about a hundred new particles.[42] André Lwoff began to study bacteriophage in *Bacillus megaterium* at the Pasteur Institute in 1949; soon Esther and Joshua Lederberg did the same in *E. coli* at the University of Wisconsin. These studies provided convincing demonstrations that lysogeny did indeed occur.

In addition to being a tool for the study of the nature of the gene, lysogeny became a model for host–virus relationships. In 1950, Lederberg asserted that "a rather simple, unitary picture of extranuclear mechanisms" could be developed if geneticists were to include the notion of "infective herdity" and agents responsible for "infective pathology" such as intracellular viruses."[43] Essentially, he offered two arguments for eradicating the distinction between an infectious particle and an integrated component of the genetic system. The first argument was based on evolutionary considerations: the cell–virus complex was "at least theoretically capable of adaptive evolution as a consequence of natural selection and mutation in the virus or the cell component or both.[44] The second reason was practical. Extrachromosomal agents which could be transferred outside the cell provided the most suitable material for experimental study of their composition, numbers, and morphological structure. "From this point of view," Lederberg remarked, "the lysogenic viruses carried by many bacteria provide the best material, especially as infection with such a virus is formally indistinguishable from events such as pneumococcus transformations."[45]

Lederberg further suggested that the cell itself could be understood in terms of interacting "quasi-organismic entities," and one of the results of seeing it this way would be that, by studying genotypic interactions from viruses to plasmagenes, geneticists could extend their models of intracellular relations to the realm of human social relations. "At each level of interaction pathological deviations can be found, ranging from sick plastids and malignant tumors (on Darlington's theory) to human serfdom."[46] Lederberg never developed this last suggestion. But he did develop the argument about the necessity of including symbionts into the genetic system of the complex organism, and he revived the idea that chloroplasts and mitochondria were intracellular symbionts. He put forth these views in 1952 in an extensive overview entitled "Cell Genetics and Hereditary Symbiosis."

The Plasmid: A Discursive Ploy

Discussions of the scope and significance of cytoplasmic inheritance had scattered through the genetics literature a plethora of terms—plasmagenes, bioblasts, plastogenes, chondriogenes, cytogenes, and proviruses—but each had acquired vague or contradictory connotations. At the risk of adding to this list, Lederberg proposed "*plasmid* as a generic term for any extrachromosomal hereditary determinant. The plasmid may be genetically simple or complex."[47] But Lederberg's plasmid was by no means an impartial term introduced to neutralize sometimes conflicting connotations of the others. It implied that all extranuclear entities would have to be treated equally as genetic constituents of cells, regardless of their origin or function.[48] It was an argument that allowed for agnosticism with regard to the origins of intracellular constituents. It was a discursive maneuver that worked in two directions: a self-reproducing cytoplasmic entity of exogenous origin could and should be treated as part of the genetic constitution of the cell, and therefore (in reverse) any part of the genetic constitution of the cell cytoplasm might be of exogenous origin. A particle shown to be infectious would have to be treated as if it were not; a particle not shown to be infectious would be treated as if it once might have been.[49]

Lederberg named infection by bacteriophage *transduction*—"a restricted transfer or genetic material to the cell"—and suggested that it was "functionally and perhaps phylogenetically, a special form of sexuality."[50] Although he viewed lysogenic bacteria as a stable symbiotic association, he remained agnostic about the ultimate origin of bacteriophages.[51] He further made a three-way association of lysogeny, genetic evidence for cytoplasmic heredity, and the many examples of hereditary symbiosis described in Buchner's *Tier und Pflanze in Symbiose* of 1932. Buchner, however, had never associated evidence for symbiosis with the view that intracellular symbiosis was a universal phenemonon. In Lederberg's view, Buchner seemed reluctant even to play up the fact that many cases of symbiosis were hereditary.[52]

Lysogeny was the workhorse in Lederberg's account, as lichens had been for those before him who had maintained that new organic wholes could by synthesized through symbiosis. The problem of delimiting the normal territory of the cell from included plasmids was magnified when the components were of equal stature. This was the case, for example, when genetically different nuclei coexist within a single cell. This phenomenon of heterokaryosis was known to persist in ascomycetous fungi. Lederberg argued that the heterokaryon permitted the association of diverse genotypes and was a mechanism for producing potentially adaptive variation, like sexual recombination, but it was less productive. Perhaps it was an early phylogenetic step in the development of sexuality.[53]

The Unbound Organism

The interdependence of organic entities in symbiotic association underscored, for Lederberg, the difficulty of delimiting the boundaries of the organism. He expressed the problem in terms of genetics. Only green plants and a few special microorganisms were autotrophic, that is, potentially self-sustaining in the absence of any organic nutrients. The rest of the living world, he argued, depended on other organisms, ultimately the green plants. In other words, he asserted, all heterotrophs were "genically insufficient" and depend "on the biosynthetically expressed genic functions of other organisms." Thus he argued that it was possible to formulate a graded series of symbioses as genic interactions, from the cohabitants of a single chromosome, through plasmids, to extracellular ecological associations of variable stability and specificity.[54]

Adopting an instrumentalist approach to what counted as an organic whole he argued that his discussion of symbiosis showed that "the delineation of organic units, be they genes, plasmids, cells, organisms, genomes or colonies is a tool of investigation and communication and not an absolute ideal."[55] From the point of view of generating new taxa, Lederberg asserted that endosymbiosis was comparable to hybridization; it was a way of bringing phylogenetically distinct genomes into intimate association.[56]

Like many of his colleagues of the 1950s, Lederberg knew well of the "old ideas" of Famintsyn and Merezhkovskii that organelles such as chloroplasts originated as symbiotic microorganisms. He had read Wallin's little book, *Symbionticism and the Origin of Species*. But for him now to express similar ideas raised the question of the attention and credibility previously withheld from these ideas. In this regard, he made the following statement:

> We should not be too explicit in mistaking possibilities for certainties. Perhaps the disrepute attached to some of the ideas represented in this review follows from uncritical overstatements of them, such as the Famintzin–Merechowsky [*sic*] theory of the phylogeny of chloroplasts from cyanophytes or the identity of mitochondria with free-living bacteria.[57]

Despite these remarks, one could never reduce the poor reception of these previous symbiosis theories to the nature of their specific formulations and rhetoric any more than one can reduce their arguments to "uncritical overstatements." In light of the polemics and disciplinary and institutional battles revealed in this study, it would be difficult not to attribute lack of positive reception of these ideas as much to the dogmatism of their detractors as to the dogmatism of their proponents.

The poor reception of earlier symbiosis theories is better seen in the light of divergent and conflicting research aims and approaches and theoretical commitments that accompanied them. After all, it was only when geneticists themselves produced evidence for hereditary symbiosis that they began to take notice of it, by associating it with viruses—naked genes. Herein lies a major difference: in previous debates, evidence for the origin of chloroplast and mitochondria was based on comparisons of morphological and physiological properties of these bodies with free-living bacteria. In the context of microbial genetics of the 1940s and 1950s, intracellular symbiosis was reduced to "plasmagene versus virus," and symbiosis—from the cell to ecological associations (and perhaps human social relations)—was reduced to genic interactions.

But there was still another important difference between Lederberg's genetic approach to symbiosis and approaches put forward by previous theorists who had argued the cell was a product of symbiosis. Lederberg explicitly denied any criteria for solving the question of plasmid origins: "The general criteria that have been used to decide the historical origin of particular plasmids are unverifiable, and such controversies have therefore tended to be sterile."[58] His own discussions illustrated "how readily the properties of plasmagenes may be imputed to viruses, and vice versa." That was enough; it was now far more constructive to focus attention on plasmid functions. Whether the behavior of extranuclear agents of heredity simulated symbionts or parasites was important, but such behavior, he asserted, was incidental to their genetic functions. Since they constituted part of the organic whole, Lederberg argued, they could not be ignored or treated as exceptions of little significance to genetics. Again, he argued that what counts as an individual organism should be defined in terms of the practices of various specialties rather than as an absolute ontological category:

> The cell or the organism is not readily delimited in the presence of plasmids whose coordination may grade from the plasmagenes to frank parasites. . . . The geneticist may well choose that entity whose reproduction is unified and hence functions as an individual in evolution by natural selection. The microbiologist will focus his interest on the smallest units he can separate and

> cultivate in controlled experiments, in test tubes, eggs, bacteria or experimental animals. Genetics, symbiotology and virology have a common meeting place within the cell. There is much to be gained by any communication between them which leads to the diffusion of their methodologies and the obliteration of semantic barriers.[59]

In the meantime, there were still relatively few genetic cases of cytoplasmic heredity as compared to nuclear heredity, and it was not certain whether or not this was simply due to a lack of adequate techniques for its detection and study. Both Sonneborn and L'Héritier found in the plasmid concept a safe shelter for *kappa* and *sigma*. In 1955, L'Héritier argued that even if the genoid was a virus it would have major hereditary and evolutionary consequences.[60] The same year, Sonneborn argued that there was no proof that *kappa* was or was not a parasite. "In Lederberg's useful terminology, it is a plasmid. Its chief present interest is as a model case on the important borderland between genetics and parasitology."[61]

"Switching" Metaphors

By the late 1950s, the claim that cytoplasmic heredity was a universal phenomenon was losing ground. Not only were very few new cases reported, but some of the previous evidence in support of cytoplasmic hereditary determinants was coming undone. Evidence previously upheld in favor of plasmagenes affecting mating type and serotype specificity in paramecia could be accounted for in terms of self-perpetuating regulatory states without recourse to the existence of cytoplasmic hereditary determinants. One seemed to be left with only *kappa*. Even the bacteriophage *lambda* which the Lederbergs had first investigated with the hope that it would prove to be a model system for cytoplasmic heredity turned out, by 1956, not to reside in the cytoplasm; it (rather its DNA) became integrated with the bacterial "chromosome."[62]

Perhaps more damaging, the theoretical necessity of invoking cytoplasmic hereditary determinants for accounting for cellular differentiation, which had been long maintained by embryologists, was being undermined. New models of cellular differentiation based on nuclear gene regulation reinforced a nucleocentric view of the cell. The idea that the nuclear monopoly resulted from a technical advantage was weakening. Perhaps the absence of evidence was actually evidence of absence after all.

During the 1940s and 1950s, genetic studies of cytoplasmic heredity had found their meaning largely in attempts to resolve what was commonly referred to as the "developmental paradox": How can cells with identical nuclear genes come to be differentiated? Those geneticists who supported a cytoplasmic basis for cellular differentiation had offered two arguments in favor of their view. First, there was no reason to believe that nuclear differentiation ever occurred, whereas they did have demonstrations of cytoplasmic differentiation and they did offer plausible models. Second, it was clear that development was an orderly and directed phenomenon; yet geneticists knew no means to induce specific gene changes.[63]

During the late 1950s and early 1960s, this debate changed dramatically. Demonstrations of nuclear differentiation began to appear, and dissent occurred among the ranks of cytoplasmic researchers.[64] Plasmagene theory as a model of cellular differentiation was supplanted by systemic conceptions of cells. Studies of nuclear regulation and feedback mechanisms culminated with the *lac* operon system in bacteria, as put forward by Jacques Monod and François Jacob at the Pasteur Institute. Their demonstrations that genes could be switched on and off in bacteria represented the pinnacle of what was often perceived as the revolution in developmental biology.[65] Cellular differences, even among cells with identical sets of nuclear genes, could be due to the activity of those identical sets of genes. In 1965, Jacob, Monod, and André Lwoff were awarded a Nobel Prize for their work on regulatory systems in bacteria.

Technological metaphors quickly replaced social metaphors. Models of cellular differentiation in terms of populations of elementary organisms responding differentially to environmental influence and struggling with each other in various modes of conflict and cooperation were replaced by communication networks, circuitry, feedback loops, and information. "The cell," wrote Jacob and Monod, "must be visualised as a society of macromolecules, bound together by a complex system of communications regulating both their synthesis and their activity."[66] In the model they described for bacteria, enzyme synthesis was regulated at the molecular level by circuits of transmitters (regulatory genes) and receivers (operators) or cytoplasmic signals (repressors) which control the rate of messages sent from the nucleus.

There was a growing belief in the early 1960s that all the creative information of the cell and organism was localized in DNA of the cell nucleus. The information contained in the DNA molecule was understood as a code, which after being transcribed into RNA in the cytoplasm was translated into protein structures. Once the information is translated, all the chemical reactions which build up the cell are also specified. Thus it was understood that nuclear genes regulate the inherited characteristics of the cell, organism, and species: the self-duplication of DNA and its regulation was the basis of heredity. However, this was not the whole story. There were serious experimental demonstrations during the 1960s opposing the idea that DNA had a monopoly on the storage of information essential for the production of structure and function.[67] But even if one assumed that all the creative information was encoded in DNA, some opted to argue that this information was not localized solely in the cell nucleus. During the early 1960s, a few biologists continued to discuss the creative aspects of hereditary symbiosis. Following the arguments of Lederberg, they maintained that there were no absolute criteria for distinguishing between structures native to the cell and symbionts.[68]

11

Molecular Reconstruction

> The pressing problems of coexistence in world affairs may have influenced the Committee in their choice of subject for this year's Symposium. If so, it is to be hoped that the more bizarre examples of symbiosis illustrated in this volume will not be followed in the world at large; there are many other ways of escaping the Hobbesian predicament that without "commonwealth" life must be "nasty, brutish and short."
>
> P. S. NUTMAN and BARBARA MOSSE, 1963[1]

There were few books chronicling the creative effects of microsymbiosis. During the 1950s, Lederberg wrote letters to several biologists and publishers seeking to have Buchner's encylopedic text translated into English. However, there were problems. A preliminary survey by one publisher did not get an "encouraging response." But the main obstacle was finding a suitable translator; it was hardly a rewarding task. As Lederberg wrote to a colleague, "It's a dog's work, and would have to be paid unhandsomely."[2] In his view, the ideal candidate would have been Paul Buchner's former professor Richard Goldschmidt. Following the appearance of Lederberg's paper on hereditary symbiosis in 1952, Goldschmidt wrote to him, commending him for mentioning Buchner's work:

> I read recently your penetrating analysis of cytoplasmic heredity. As usual among weak humans, I liked best the parts where my own thinking led me in similar directions. I was specially pleased that you, as probably the only US geneticist, are aware of the great theoretical importance of my friend and former student Buchner's work on symbionts, generally unknown also among zoologists.[3]

Goldschmidt agreed that Buchner's book should be translated. "As a matter of fact," he wrote Lederberg, "I think it's a disgrace that the run of American biologists have never heard of the outstanding morphological work of the last generation with all its general consequences." Unfortunately Goldschmidt could not think of doing the translation himself since he was in the midst of writing his own book and was ill with heart trouble and still working at home.[4] A translation of Buchner's book did not appear until 1965.[5] Nonetheless, the classical research on microsymbiosis was left in the dust of those who rushed to exploit the new field of molecular biology. Symbiosis itself would have to be reconstructed from the bottom up before it attracted the interest of mainstream biology in the United States.

The lack of interest in symbiosis per se should not be taken to imply that there was a lack of research pertaining to symbiosis. But such work continued to be scattered and divided among diverse specialties with divergent interests, mainly practical, and the literature itself was dispersed in specialist journals. Long-established institutional barriers kept studies of symbiotic relationships apart. The microbiologist Clark Read pointed to these problems in 1958 when calling for an organization of efforts pertaining to studies of symbiosis in the United States. Various kinds of parasitism from worms to viruses were studied primarily from medical, veterinary, and agriculture standpoints. Those interested in plant relationships with other plants, animals, or microbes were originally identified with botany, while those who worked with animals that live in and on animals had their origins in zoology. Bacteriologists, on the other hand, were primarily concerned with studies of bacteria in vitro. As Read saw it, biologists knew little more about the behavior of bacteria in vivo than in the days of Pasteur.[6]

Read called for the establishment of a department or institute in some major university devoted to the physiological basis of symbiosis, with a special committee empowered to organize courses, sponsor research, and grant degrees in symbiosis studies. Rather than representing a new specialized discipline, such a new department, he maintained, would represent "a combination of disciplines, a symbiosis, having an aim which is synthetic in nature." The benefits of such an interdisciplinary program would be felt in many areas not directly concerned with symbiosis studies per se: "The area of biological control, problems of world nutrition, population manipulations," and like Lederberg before him, Read suggested that "perhaps even some areas of direct human relationships might well be better understood."[7]

Poor Cousins

Read's vision of an interdisciplinary symbiosis institute was never realized. The creative nature of symbiotic associations continued to be downplayed in biology texts. Parasitology and ecology texts often included a brief section on the more commonly used examples of mutualistic symbiosis, leaving the impression that it was an exceptional condition in nature.[8] General biology, zoology, and botany textbooks usually did little more than mention the topic. However, a few popular articles on specialized cases and popular books on symbiosis began to appear in the late 1950s and 1960s, often describing the relations in terms of cooperation.[9] In the more numerous technical papers on microsymbiosis, mutualism was usually understood in terms of a bilateral exploitation.[10]

The creative effects of symbiosis also continued to be overshadowed by overwhelming attention to illustrations of conflict and competition—a view which, it had long been argued, only reflected dominant views of human social progress. Perhaps it is not surprising that the first major conference on symbiosis, with botanists, zoologists, and microbiologists coming together to discuss the different types of association, was not held until April 1963, hosted by the Society for General Microbiology, in London. These were extraordinary times indeed, at one of the confrontational peaks of the Cold War: the Cuban missile crisis. That the Society for General Microbiology hosted a meeting on symbiosis at this politically troubled time appeared to be no coincidence, as P. S. Nutman and Barbara Mosse, the editors of the resulting volume, *Symbiotic Associations,* suggested.[11]

The editors recognized that the subject of symbiosis was usually considered to have no "independent standing" and that "its name tends to disguise a fundamentally parasitic relationship," masking it with an "unscientific aura of teleogy." They insisted, however, that while symbiosis could not be used to invoke examples of purposeful cooperation, its beneficial effects could not be simply undermined, overlooked, and understood solely in terms of parasitism. They argued that the Hobbesian, all-against-all reasoning could be reversed, and that parasite–host relationships might be better understood from the standpoint of implicit cooperation, rather than the reverse.[12]

The emphasis on the negative effects of symbiotic associations was, of course, by no means simply a reflection of Western ideology. It was reinforced by medical constructions of the inherent pathological nature of bacteria and viruses. The few stories about the creative and adaptive effects of miscrosymbiosis had been no match for those which emphasized the war between humankind and microbes.[13] The "beneficial" effects of microbial symbiosis continued to stand in virtual conflict with the very history that had brought science and medicine into intimate association.

The organizing committee for the Symposium on Symbiotic Associations directed participants to focus attention on those "*symbiotic relations involving microorganisms in which both members derive ecological advantage.*" Bacteriologists René Dubos and Alex Kessler, both of the Rockefeller institute for Medical Research, gave a theoretical introduction elaborating on the creative dynamics of symbiotic relationships. They argued against the view which reduced the outcome of relationships to the inherent properties of the organisms themselves. It was not simply that certain kinds of organisms were parasites and others were mutualists. The performance of associations was dependent on environmental factors that operated at any one given moment. "For this reason," they explained, "the terms most commonly used to denote biological relationships—such as symbiotic, commensal, mutualistic, parasitic or even pathogenic—are rather misleading if they imply a state of permanency of the relationship which does not take into account its dynamic character."[14]

During the late 1950s and early 1960s Dubos had become well-known as one of the few advocates for the integrative and creative effects of infection. Symbiosis was virtually absent in his 1945 book *The Bacterial Cell,*[15] but by

1960 he could list various well-established examples drawn from research over the previous decade which showed how the "interplay of biological systems" could result in the appearance of new structures, functions, properties, and products: the discovery that diphtheria was due to the infection of the bacterium involved with a toxogenic virus; symbiotic phage brought about profound changes in the anatomy of *Salmonella; lambda* phage carrying the *gal* region derived from bacterial genes of a host cell confers upon another recipient bacterium the ability to produce enzymes for the utilization of galactose.[16] He added these examples to the classic cases of root nodules of legumes, the morphogenetic effects of the association of algae and fungi in constructing lichens, and to characteristic tumors (crown galls) induced by inoculating certain plants with a certain bacteria.[17]

With these cases in hand, at various symposia on virology and microbiology, Dubos turned his polemics against the myopic views of his fellow microbiologists and virologists who continued to study and understand infection exclusively in terms of pathology. The dominant preoccupations of participants at symposia on virology during the 1960s centered around such themes as the origin of poliovirus constituents and, especially, virus-induced cancer. But Dubos asserted that the establishment of a potential pathogen in the host rarely manifested itself in the form of disease. In many cases, he declared, viruses might actually enhance the host's ability to survive and thrive under certain ecological conditions.

The examples of the integrative and creative effects of infections, as well established as they were, still could not tip the balance against the mass of evidence for the pathological nature of microbial and viral infections. Dubos could only reply, as others had before him, that this evidence was not an image of nature; it only reflected the research interests shown by virologists and microbiologists. Searching for integrative processes was not "a scientifically fashionable occupation." He chided virologists and microbiologists that despite their claims of independence, most still behaved as "poor cousins in the mansion of pathology." They were eager to show how the phenomena they studied were associated with pathology, where funds could be obtained, and not with general problems of biology. If the creative effects of infection were studied as intensively as were its pathological effects, Dubos prophesied, "there would soon develop a new science of cellular organization, indeed perhaps a new biologic philosophy." The time had come, he declared, "to supplement the century old philosophy of the germ theory of disease with another chapter concerned with the germ theory of morphogenesis and differentiation."[18]

The Genetics of Symbiosis

Emphasizing the integrative and creative manifestations of symbiosis, at the Symposium on Symbiotic Associations in London in 1963, Dubos and Kessler raised the possible relevance to cell origins. It was well known that one could remove from certain cells various organelles, provided the cell could call into play alternative metabolic pathways. This had been shown for "petite mutations" affecting mitochondria in yeast. The unicellular alga *Euglena* could also be rid of chloroplasts by treatment with the antibiotic streptomycin. To Dubos and Kessler these facts clearly indicated that cells had several mechanisms of information storage, that one could think of the cell not as a genetic unit but rather as a symbiotic assembly of several independent genetic organelles which have become thoroughly integrated.[19]

With the exception of one paper on lysogeny, the Symposium on Symbiotic Associations focused largely on analyses of the classical cases: lichens, bacteria in root nodules of legumes, mycorrhiza. But Lederberg's plasmid concept was used to embrace symbiotic algae in invertebrates.[20] It was especially useful to those protoozologists who investigated symbiotic algae in paramecia, such as Stephen Karakashian at the University of California. For him, it implied that the paramecium–chlorella association represented a certain initial stage in the mutual adaptation of host and plasmid and therefore possessed an evolutionary potential for considerably more integration.[21]

In 1965 Karakashian and R.W. Siegel attributed three virtues to what they called "the genetic approach to symbiosis."[22] First, it had synthetic powers: it provided a conceptual framework around which to organize a mass of diverse data from bacteriophages to bacteria on the root nodules of legumes. For example, the ability of rhizobium-infected root nodules to fix nitrogen and the appearance of new antigens on the surfaces of virus-infected bacteria both illustrate new phenotyes arising from acquisition of a symbiont; endosymbioses therefore "either create or release genetic information." Second, it emphasized the genetic and biochemical integration of host and symbionts. Together they form a single biological system, more or less perfectly coordinated, so that distinctions between "invader" and "organelle" may become impossible, or at least meaningless. Third, the approach had heuristic value. It gave rise to new questions. For example, what factors regulate the respective division rates of host cells and symbionts so that neither outstrips the other and the association is faithfully perpetuated? How complex is the genetic control of mutually compatible host and symbiont specificities, and how do these genes operate? Are macromolecules exchanged between symbiont partners, or was macromolecular synthesis necessary for the creation of a suitable intracellular environment for symbiosis? Are new molecular species formed as a result of the juxtaposition of two dissimilar genomes? These kinds of questions could be answered by applying genetic and biochemical techniques to the biology of host–symbiont relationships.[23]

However, not all of those who studied symbiosis believed that the genetic approach was going to lead to a new view of cell structure and function. Classical symbiosis researchers often seemed to be immune to the new attitudes and approaches that were forming. In 1966 and 1967, S. Mark Henry edited two major volumes on symbiosis bringing together recent work of American and European biologists. However, the idea that all cells were symbiotic complexes was discussed in one paper, and then only to be ridiculed.[24] Paul Buchner remained adamantly opposed to it. In the revised English translation of his text *Endosymbiosis of Animals with Plant Microorganisms* (1965) he discussed the symbiotic origin of organelles under the heading "Wrong Paths in Symbiosis Research."[25] Dissociating his own work from the influence of such theories, he wrote, "For us who have remained aloof from such speculations, endosymbiosis . . . represents a widespread, though always supplementary device, enhancing the vital possibilities of the host animals in a multiplicity of ways."[26]

Eukaryote Origins: The Space Race

When referring to the speculative nature of such theories Buchner expressed a common sentiment among biologists who believed that such origin stories were simply not scientific, in the sense that one could not prove or disprove them. Yet other events in the Cold War and expectations of new genetic and biochemical approaches led to an appreciation of just such origin accounts, and during the late 1950s and early 1960s studies of the origin of life began to appear in abundance.

On 4 October 1957, *Sputnik* was launched. The possibility of space exploration spurred scientists' interests in the problem of ultimate origins—the origin of the universe and the origin of life—and the study of life on other planets, which in 1960 Lederberg dubbed "exobiology."[27] Lederberg himself had been in Australia working with MacFarlane Burnet at the University of Melbourne and witnessed the ascent of *Sputnik* in the southern skies, an event that filled him with both awe and apprehension. On returning to his own laboratory, he read extensively on astronomy and rocketry, and by December he had sent off memos to several influential scientists asking for their help to avert what he saw as a potential "cosmic catastrophe": the contamination of life forms that might be present on other planets by organisms carried from earth via space flight. As he recalled, "My fear was that the space program would be pushed ahead for military and political reasons without regard for the scientific implications."[28] Thanks to the efforts of Lederberg and a few influential colleagues who were officers of the National Academy of Sciences, an international committee was formed to establish guidelines and plan methods for detecting and protecting life in space. Lederberg served on the academy's committees on space biology from 1958 to 1977 and on NASA's lunar and planetary missions boards, involved with the *Mariner* and *Viking* missions, from 1960 to 1977.[29] Though the Mars *Viking* mission did not find life on Mars, such interest provided an intellectual climate more tolerant to speculations on the early evolution of life.

While new technologies were being developed for viewing outer space and funds from NASA were provoking scientists' interest in the origin of life, other microchemical technologies had been developed for viewing inner space. During the 1940s and 1950s, easy to operate and reliable electron microscopes had become available that gave resolution of less than 3 angstroms. Such high resolutions coupled with suitable advances in the methods of specimen preparation permitted the direct visualization of large molecules.[30]

Once cell structure could be visualized at magnifications many times those achievable with a light microscope, it became easy to define the similarities of the smallest cell types and to distinguish them from other unicellular microorganisms. In the late nineteenth century the general term "bacteria" had been used to embrace those minute, rounded, ellipsoid, rod-shaped, threadlike, or spiral forms of "vegetable organisms" sometimes spoken of as belonging to the class of lower fungi.[31] Though bacteria and blue-green algae had often been regarded as "primitive cells," there was no consensus regarding the differences between these cells and other cells. This problem had remained at the center of debates over the bacterial nature of organelles since the late nineteenth century. Studies of the internal organization of cellular machinery through the electron microscope combined with new biochemical techniques[32] quickly led to a revival of interest in whether all plant and animal cells were symbiotic complexes.

In 1962, R. Y. Stanier and C. B. van Niel at the University of California, Berkeley, and the Hopkins Marine Station of Stanford University maintained that among all organisms there were only two different organizational patterns of cells, those which the French protozoologist Edouard Chatton had in 1937 distinguished as the *eukaryotic* and *prokaryotic* types.[33] Bacteria and blue-green algae could be clearly distinguished from all other protists (other algae, protozoa, and fungi) on the basis of the prokaryotic nature of their cells.[34] Unlike prokaryotes, eukaryotes possessed internal membranes which separated the nucleus from the cytoplasm. Mitosis occurred only in eukaryotes.[35]

Within the cytoplasm of eukaryotic cells the enzymatic machinery for photosynthesis and respiration was located in specific organelles: chloroplasts and mitochondria. Indeed, since the 1950s there had been a great deal of biochemical research on mitochondria. They were no longer considered to give rise to all sorts of cell structures, as the eclectosome theory had proposed. They were the seat of respiratory enzymes and the energy transduction system. An important part of mitochondrial research was concerned with the mechanism of electron transport and oxidative phosphorylation.[36]

No such metabolic units existed in prokaryotes. Photosynthesis and respiration in prokaryotes could only be attributed to the cell as a whole, considered as a metabolic unit. Only the eukaryotic cell appeared to have the potentialities for the development of highly differentiated multicellular biological systems, and only this kind of cell was perpetuated in the evolutionary lines which eventually gave rise to higher plants and animals.[37] The evolutionary implications of the distinction between prokaryotes and eukaryotes were profound. In their book *The Microbial World* (1963), Stanier, Douderoff, and Adelberg remarked, "In fact, this basic divergence in cellular structure, which separates the bacteria and bluegreen algae from all other cellular organisms, probably represents the greatest single evolutionary discontinuity to be found in the present-day world."[38] The evolutionary origin of the eukaryotic cell became a subject of considerable interest.

The other major impetus for the renewed interest in symbiotic theories of the origin of cell organelles arose out of further evidence for the existence of cytoplasmic genes and gene complexes in all organisms. In view of the demonstrations that genetic systems in viruses and in nuclear elements of cells were associated with DNA or RNA, several biologists since the mid-1950s had looked for nucleic acids in plastids. But the presence of DNA in plastids remained debatable. The bugaboo was the ever-present problem of contamination, this time not microbial but nuclear. One needed an organism in which suitable staining techniques could be used, avoiding such problems of contamination. The unicellular green alga *Chlamydomonas* was found to be most useful.[39] *Chlamydomonas* emerged as the model organism for genetic analysis of chloroplasts with the now classic researches of Ruth Sager and Nicholas Gillham during the 1960s and 1970s.[40] Genetic studies of mitochondria soon followed.

When new, clean—that is, noncontroversial—demonstrations of DNA associated with cytoplasmic organelles were published in the early 1960s, the question of symbiosis was placed neatly on the tip of the tongue of authors discussing such evidence. When in 1962 Hans Ris and Walter Plaut, at the University of Wisconsin, provided evidence for chloroplast DNA in *Chlamydomonas,* they pointed to a striking likeness between DNA-containing regions in chloroplasts and the nucleoplasm of blue-green algae:

> We suggest that this similarity in organization is not fortuitous but shows some historical relationship and lends support to the old hypothesis of Famintzin (1907) and Mereschkowski (1905) that chloroplasts originated from endosymbiotic bluegreen algae. . . . With the demonstration of ultrastructural similarity of a cell organelle and free living organisms, endosymbiosis must again be considered seriously as a possible evolutionary step in the origin of complex cell systems.[41]

The next year (1963) when Sylvan Nass and Margit Nass at Stockholm University published their evidence for DNA in mitochondria, they put forward the symbiotic nature of mitochondria and its identity with bacteria:

> We have already stressed the similarities between mitochondrial fibers and bacterial nucleoplasm with regard to their structural appearance and their fixation and stabilization properties. There is a great deal of modern biochemical and ultrastructural evidence that may be interpreted to suggest a phylogenetic relationship between bluegreen algae and chloroplasts, and bacteria and mitochondria.[42]

The Argument from Dr. Pangloss

Although the possibility that chloroplasts and mitochondria are/or were symbionts gained interest and plausibility from these results, not all biologists found the argument to be compelling. In 1967, Richard Klein and Arthur Cronquist of the New York Botanical Garden rejected the notion that chloroplasts are symbionts as a "bad penny" [that] has been circulating for a long time."[43] The idea that organelles originated as symbionts had been discussed at length the previous year when the International Society for Cell Biology hosted a symposium on "The Formation and Fate of Cell Organelles." Some left the question open as to whether or not "a mitochondrion is the overfed remains of a symbiotic bacterium," and focused solely on the structural and biochemical similarities.[44]

Aharon Gibor from the Rockefeller University in New York assessed the strength of the evidence for and against the argument that organelles were well-integrated symbionts that invaded a primitive cell early in evolution. He found four sources for the appeal of this hypothesis: (1) different degrees of integration of a symbiont into the life of a cell were known to occur; (2) the DNA of the plastids and mitochondria differ qualitatively from the nuclear DNA; (3) the plastids and mitochondria are separated by a double membrane from the rest of the cell; (4) the organization of the DNA fibers of the organelles, as revealed by electron microscopic studies, resembles the DNA organization in a bacterial nucleoid rather than the chromosomes of the nucleus of higher organisms.

On the other hand, Gibor listed a number of problems for the symbiotic theory: (1) the fact that the organelles of different species differed in their DNA suggested that different symbionts would have had to enter different cells and all would have proceeded to evolve in an identical pattern; (2) some of the enzymes of the organelles are coded for by nuclear genes and are probably synthesized in the cytoplasm and then integrated into the organelles; (3) some of the RNA of the mitochondria appeared to complement nuclear DNA, suggesting that it had been synthesized in the nucleus and transferred into the mitochondria.[45]

In further support of an endogenous origin of such organelles, Gibor argued that there were two "obvious advantages" to the cell to having some of its DNA localized in cytoplasmic organelles: "Ontogenetically, the cell is provided with an efficient and rapid mechanism for the synthesis of essential proteins; phylogenetically the multiplicity of templates can function to stabilize very important enzyme systems from the hazards of deleterious mutations."[46] Behind these statements is the argument that since one could find a good fit between structure and function, the cell itself would have evolved this way for this reason: the last thing a cell or organism would "want" is a mutation affecting fundamental mechanisms crucial to its survival. The geneticist Ruth Sager expressed this view with precisely this teleological language in 1966: "The existence of a nonchromosomal genetic system designed to minimize variability leads one to wonder whether NC genes control particular traits of crucial value to the organism."[47]

This good fit of form and present function permitted in itself no conclusion about how these organelles or any other organismic structure originated. Nonetheless, this quick move between current function and reason for historical origin has always been prevalent among evolutionists, especially among those advocating gradual evolution, neo-Lamarckians and Darwinians. In 1914, William Bateson, himself a staunch opponent of gradual evolution, argued against precisely this kind of reasoning when he wrote: "Naturalists may still be found expounding teleological systems which would have delighted Dr. Pangloss himself, but at present few are misled."[48] Since Bateson's pronouncement, evolutionists sensitive to this fallacy have continued to refer to Voltaire's satire and have called upon the venerable Dr. Pangloss to expose it: "Things cannot be other than they are, for since everything was made for a purpose, it follows that everything is made for the best purpose. Our noses were made to carry spectacles, so we have spectacles. Legs were clearly intended for breeches, and we wear them."[49] None of this is to say, of course, that the structures in question could not have been constructed by natural selection for their current "purpose." If organelles did evolve gradually in the classical Darwinian way, then one might expect to find transitional forms in some bacteria, in the same way one could find examples of various stages in the integration of a symbiont into the cytoplasm of some eukaryotic cells. In fact, Gibor could point to just such evidence.

In a paper on "The Biogenesis of Mitochondria," published in *Nature* in 1966, D. Halder, K. Freeman, and T. S. Work at the National Institute for Medical Research in London argued that a specialized region of the bacterial membrane, "the mesosome," had the biochemical character of a primitive mitochondrion and seemed to combine the roles of mitochondria and centrioles in higher cells. They were "forced to suggest that the mesosome, which had the capacity to attach to the bacterial chromosome, may well have acquired a few of the chromosomal genes and subsequently differentiated further into the various related structures of higher forms, namely, the centrioles, the mitochondria, the kinetoplast, the basal bodies and the plastids."[50]

Serial Endosymbiosis Theory

The next year, another paper on the "Evolution of Eucaryotic Cells" appeared in *Nature*. It was by the Norwegian microbiologist Jostein Goksøyr at the Institute for General Microbiology, University of Bergen. He argued against the view that one prokaryotic cell developed into a eukaryotic one, and he constructed a series of "logical steps" whereby one eukaryotic cell could have evolved from a number of prokaryotic cells. In a first step, anaerobic prokaryotes of a single species were brought into contact without intervening cell walls. The DNA of the individual cells accumulated in the center of the compound cell and a primitive mitotic mechanism developed. A nuclear membrane developed from endoplasmic reticulum as the mitotic process developed further. The resulting cell was then an anaerobic eukaryote.

Aerobic prokaryotes emerged later when phytosynthetic blue-green algae produced oxygen in the atmosphere. Subsequently, some of the anaerobic prokaryotes established an endosymbiotic relationship with the aerobic forms.

During further evolution the aerobic partner lost a great deal of its autonomy when some of its DNA became incorporated in the nuclear DNA, "giving the eukaryotic cell a still better control over its aerobic partner." The final step in this evolutionary process was the development of mitochondria as they occur today. The aerobic eukaryote would enter a new symbiotic relationship with a primitive kind of blue-green algae which developed along the same lines as mitochondria to appear as the chloroplasts of eukaryotic algae and higher plants.[51] Unlike Gibor, Goksøyr found no problem with the suggestion that such symbiotic systems could have developed a number of times from different prokaryotic forms.

Goksøyr did not believe that his logical steps could ever be proven, but he suggested that one might carry out DNA/RNA hybridization experiments between blue-green algae and the chloroplasts of other algae. The assumption underlying this suggestion is that the longer two organisms have been separated from a common ancestor, the more base sequence changes will have occurred in each of their genomes, and therefore the less similarity there is between their DNAs. The extent of pairing of nucleotide sequences thus could be used to infer phylogenetic relationships of organisms. This new molecular biology technique became one of the principal ways to establish the phylogenetic relatedness of cytoplasmic organelles to each other and to the nucleus of their "host."

The same year that Goksøyr produced an outline, Lynn Margulis (Sagan) published a general hypothesis for the origin of eukaryotic cells by a succession of endosymbioses.[52] Her detailed, provocative, and stimulating work—including her now classic text, *Origin of Eukaryotic Cells* (1970) and the first edition of her *Symbiosis in Cell Evolution* (1981)—soon led to a great deal of comparative ultrastructural, biochemical, and geological work on the question. She became well recognized for constructing a coherent, detailed framework for what F. J. R. Taylor dubbed the "serial endosymbiosis theory" (SET).[53] Very briefly, the serial endosymbiosis theory holds that eukaryotic cells emerged from the multistaged establishment of symbiosis among the prokaryotes. It focused primarily on three organelles: mitochondria, chloroplasts, and the basal bodies or centrioles, asserting that these organelles originated as free-living bacteria in symbiosis with some host that became the nucleus-containing cytoplasm of the resulting eukaryote. To most supporters of the serial endosymbiosis theory, the only case which remained in doubt was the basal bodies or centrioles.

Margulis first learned of symbiosis in the late 1950s when she was a student in Ris's classes at the University of Wisconsin.[54] Students read the classical literature, including E. B. Wilson's *The Cell in Development and Heredity* and his statement about symbiosis and "polite biological society." She became interested in plastids and the idea that they were self-duplicating. Margulis completed her doctoral dissertation at the University of California, Berkeley, in 1965 on studies of chloroplast autonomy in *Euglena.* She continued her studies as a postdoctoral student at Brandeis University, where she continued to read about bacteria and microbial classification, about symbiosis, plastids, and mitochondria as well as the mitotic apparatus, or "division center" of the cell—the centrioles.

By that time there were already reports that mitochondria and chlorolasts had DNA. There were also claims that traces of DNA could be found in centrioles or basal bodies.[55] However, this evidence was controversial, as it continues to be today.[56] Nonetheless, since the 1880s several leading cytologists and protozoologists had considered centrioles to be self-reproducing organs. In the early 1950s, André Lwoff had identified centrioles with kinetosomes (basal bodies), at the base of cilia, which he believed played the predominant role in the morphogenesis of ciliated protozoa. Subsequent electron microscopic studies confirmed that there was a structural homology among these cell structures. Each was composed of a pinwheel of nine outer microtubules surrounding two inner microtubules. Much of Margulis's theory rested on accounting for these 9 + 2 homologs. She explained it by invoking a motile prokaryotic endosymbiont, a spirochetelike organism, as the common ancestor of the complex flagellum of eukaryotic cells and organs concerned with mitosis.

In 1964 L. R. Cleveland and A. V. Grimstone reported that the protozoan *Mixotricha paradoxa,* found in the gut of a "primitive" termite, *Mastotermes darwiniensis,* lived in close and constant association with at least three distinct kinds of microorganisms which were integrated into its structure and behavior. They identified structures previously taken for cilia as spirochetes, and other structures previously taken for basal granules as adherent bacteria: "*Mixotricha* is propelled not by its flagella but *by the undulations of its attached spirochaetes.*"[57] Cleveland and Grimstone did not believe that the utilization of spirochetes as a method of locomotion was of general significance for protozoa. On the contrary, based on the continual and undirected movement spirochetes seemed to provide, they suggested that this method of locomotion could have been evolved only in a sheltered environment such as that provided by the termite gut, in which it is unnecessary to search for food and avoid predators and unfavorable conditions.[58] Nonetheless, the morphological similarities between locomotive organs and spirochetes were striking, and Margulis suggested that perhaps a spirochete-like organism had been ingested by a primitive amoeboid cell.

Margulis's paper "On the Origin of Mitosing Cells," published in the *Journal of Theoretical Biology* in 1967, was by far the most daring and serious effort in pursuit of the symbiont hypothesis of the 1960s. She explored the geochemical and paleontological record as well as the data of cytology, microbiology, and biochemistry to seek relevant data.[59] Her story began with the existence of prokaryotic cells containing DNA, which she believed arose between 4.5 and 2.7 billion years ago. She postulated that eukaryotic cells arose between 0.5 and 1.0 billion years ago, perhaps a billion years after the evolution of oxygen-forming blue-green algae. The presence of atmospheric oxygen permitted the evolution of aerobic bacteria (a protomitochondrion), which were ingested into the cytoplasm of heterotrophic anaerobic amoeboid cells. The ancestral eukaryote represented the fusion of an aerobic bacterium with another prokaryote to give rise to a mitochondrion-containing aerobic amitotic amoeboid organism. Some of these amoeboids subsequently ingested certain motile prokaryotes (a spirochete-like organism). Eventually these too became symbiotic in their hosts, forming a motile ancestral amoeboid organism containing the crucial 9 + 2 flagellar symbiont, and in which classical mitosis evolved. The endosymbiont genes which determined its characteristic 9 + 2 substructure were eventually utilized to form the chromosomal centromeres and centrioles.

Margulis argued that at least two series of mutational steps were required for the endosymbiont DNA to differentiate into centrioles.[60] One led to the development of some attraction between the nucleic acid of the host (chromosomes) and that of the symbiont (centrioles). The other led the endosymbionts to opposite poles of the host cell. This process had occurred at various times and in various primitive amoeboflagellates, and these cells evolved to form protozoa or fungi or eventually multicellular animals. The evolution of mitosis, ensuring an even distribution of large amounts of nucleic acid at each cell division, took millions of years.[61] During the course of the evolution of mitosis, various lines of protozoans were infected with a blue-green alga to generate the ancestral eukaryotic alga on the way to the formation of higher plants.[62]

On the basis of her symbiosis theory Margulis made two major innovations in the classification of microorganisms in 1967. First, she separated lower eukaryotic algae into groups based on homologies of the host or protozoan part of the organism and offered this classification as a basis for resolving one of the long-standing problems in the classification of unicellular organisms. The problem of classifying phytosynthetic, motile eukaryotic organisms had been discussed since the 1860s, when questions of "animal chlorophyll" were first addressed. Were such organisms plants or animals? One hundred years later the problem had not yet been resolved. Botanists classified organisms with plastids in them as plants, while zoologists placed the same organism in their respective protozoan groups. However, if plastids were symbionts that had been acquired by different protozoan hosts, Margulis reasoned, both botanists and zoologists were right: one had to classify both components of the symbiosis separately, as was recognized in lichen classification. "Perhaps, then," she declared, "a natural phylogeny of the lower eukaryotes could be developed which is satisfactory to botanists, zoologists and mycologists."[63]

Second, she classified groups of protozoans according to their mitotic cytologies, and hence the stages they presumably represent in the evolution of eumitosis. She based this classification on her theory of the multiple symbiotic origin of the organelles concerned with mitosis. "If the theory presented in this paper is correct," she wrote, "all eukaryotes should ultimately be classified completely, correctly, and consistently according to their position in the origin of mitosis."[64] The difficulties of fitting microorganisms into the age-old categories of animals and plants later led her to link her work with the proposal of the ecologist Robert H. Whittaker, who in 1969 postulated the existence of five kingdoms: Monera, Protista, Fungi, Plantae, and Animalia.[65]

Margulis concluded her 1967 paper with the prediction that one day it might be possible to extract centrioles, mitochondria, and chloroplasts and culture them in vitro.[66] During the following decade, however, the idea that one could extract mitochondria and chloroplasts from the cell and grow them in culture rapidly faded. To the extent that the endosymbiont hypothesis rested on such claims of organelle autonomy, it began to lose ground almost as quickly as it had gained it.

12

The Dull Edge of Ockham's Razor

> It might have happened thus; but we shall surely never know with certainty. Evolutionary speculation constitutes a kind of metascience, which has the same intellectual fascination for some biologists that metaphysical speculation possessed for some medieval scholastics. It can be considered a relatively harmless habit, like eating peanuts, unless it assumes the form of an obsession; then it becomes a vice.
>
> R. Y. Stanier, 1970[1]

During the 1970s and 1980s the symbiotic theory for the origin of cell organelles, once so out of harmony with accepted theory and "too fantastic to mention in polite biological society," emerged not only as a respectable subject of scientific discourse, but as a thriving research program. This is not to suggest that all biologists readily accepted it. On the contrary, the symbiotic theory was evaluated from various perspectives and received with various degrees of agreement and skepticism. Many welcomed the argument for mitochondria and chloroplasts but rejected it for the organelles concerned with mitosis and motility. Still others were reluctant to accept the claim that chloroplasts and mitochondria evolved as symbionts. During the late 1960s and 1970s the serial endosymbiosis theory continued to be developed in direct conflict with autogenous theories labeled "direct filiation," which held that organelles evolved by compartmentalization inside cells. The nature of the debates which ensued highlight the difficulties of establishing the validity of any evolutionary account.

The Argument for Simplicity

Most of the controversy hinged not on the authenticity of the facts per se, as in previous episodes in this history, but rather on divergent judgments as to which of the alternative theories best explained the facts. This in turn led to discussions of the technical and formal criteria for the assessment of theory, what a scientific theory was supposed to do. Most discussants realized that more than one theory could account for the facts. They also recognized that one had to consider well how the corpus of facts was selected so as to conform with a given theory. Since several explanations were consistent with any set of observations, biologists upheld "simplicity" as a criterion for deciding between them.

There were two sides to simplicity. One reflected the view that nature obeys a principle of least action. August Weismann, it will be recalled, had appealed to this "economy of nature" when arguing against the idea that there would be more than one bearer of heredity in the germ plasm. But in the debate over the origin of cell organelles, simplicity was most often expressed as an epistemological maxim—that one should chose the "simplest" hypothesis fitting the facts—a maxim attributed to the fourteenth-century philosopher William of Ockham. This principle of scientific frugality became known as Ockham's razor: "Neither more, nor more onerous, causes are to be assumed than are necessary to account for the phenomena."[2]

By the late 1960s, the remarkable cytological and biochemical similarities between eukaryotic cell organelles and prokaryotes—their general structural similarity, their similar modes of multiplication by fission, the composition of their DNA—were well recognized. But critics insisted that none of this evidence could be taken as direct support for symbiotic theories. In 1969, after listing the similarities between prokaryotes and chloroplasts and mitochondria, A. Allsopp, at the Department of Cryptogamic Botany, University of Manchester, declared: "There is now, however, a wide, in the present author's opinion mistaken, acceptance of a symbiotic origin of these organelles of the eukaryotic cell, a view which is certainly supported by the widespread occurrence of microorganisms as endosymbionts."[3]

Allsopp dismissed Margulis's argument that the whole mitotic apparatus, including the centriole and centromeres, as well as the flagellar apparatus of eukaryotes, was originally derived from a flagellate prokaryotic endosymbiont, stating that there was simply little evidence in favor of "this interesting but improbable hypothesis."[4] In his opinion, all the striking similarities between organelles and prokaryotes were rather precisely what one would expect if eukaryotes had evolved by gradual transformation of prokaryotes. It was "plausible that in simple conservative structures like plastids and mitochondria, the original prokaryotic DNA, and indeed the ribosomes, should remain relatively unchanged, whereas nuclear DNA would inevitably be transformed during progressive evolution of the cell."[5]

"Theoretically," Allsopp, remarked, "there is no great difficulty in bridging the gap between prokaryotic and eukaryotic cell structure although extant transitional forms are unknown at present."[6] There was, of course, evidence for transitional stages of endosymbiosis, but Allsopp denied the generality of the phenomenon. The occurrence of such microorganisms as endosymbionts merely demonstrated that a symbiosis of type could occur "as a sporadic, if fairly frequent, anomaly." He concluded that, on the general principle that the simpler hypothesis was always to be preferred, the direct transformation of prokaryotes was a more reasonable hypothesis than the symbiotic view.[7]

The next year Allsopp's claims that the "direct transformation" hypothesis was simpler and superior to the symbiotic theory was denied by the Stanford biologist Peter Raven in Science. In his opinion the impressive homologies between mitochondria and plastids on the one hand and prokaryotic organisms on the other made it "almost certain" that these organelles had their origin as independent organisms. He interpreted the evidence for transitional stages of symbiosis as providing further support.[8] After listing the names of twenty-six authors who supposedly supported endosymbiotic origin of plastids and mitochondria, Raven concluded that in the face of contemporary evidence, the arguments necessary to defend an endogenous view were "far more complex than those supporting what is now clearly the majority opinion."[9]

The symbiotic theory may have been the majority view by 1970, but reaching consensus about which theory was the simplest was not straightforward, as John Hall commented in a letter to the *Journal of Theoretical Biology* in 1973. In his opinion the symbiotic theory was "the more plausible one, providing the simplest explanation of the known mitochondrion." Nonetheless, after noting that Allsopp had considered the same data and applied the same criterion but reached the opposite conclusion, Hall remarked that "Occam's razor is none too sharp a blade when applied to a Gordian knot of such complexity."[10]

Simplicity had problems because, on any index, there still seemed to be opposing explanations of equal simplicity. But new arguments continued to be made in the debate. The claim that the symbiotic theory was the simplest was countered by Rudolf Raff and Henry Mahler in 1972 and 1975. They understood the stakes: if the symbiotic theory were correct, then it would explain "the greatest evolutionary discontinuity between living organisms"—that separating prokaryotic from eukaryotic cells.[11] As they saw it, the symbiotic theory was "esthetically pleasing," but it was not "compelling." They recognized that there were acute difficulties in obtaining conclusive proof for any account of the origin of the eukaryotic cell and its organelles.

The events one was trying to reconstruct took place over a billion years ago. Trying to obtain relevant evidence was difficult; it was even more difficult to interpret. The fossil record was virtually absent. One had to find clues by unraveling the nature of contemporary systems. But the sheer masses of morphological, ultrastructural, ecological, taxonomic, biochemical, and genetic data available that one had to sift through were almost everwhelming. The question came down to which clues, and what evidence, were relevant. The evidence one selected and emphasized depended upon the theory and the discipline one favored. The apparent strength of the symbiotic theory, in Raff and Mahler's view, resulted from these biases. Most of the evidence presented in favor of the symbiotic hypothesis had been drawn from cytological studies of cells and from comparative studies of living intracellular symbionts, organelles, and free-living prokaryotes. But Raff and Mahler concentrated on comparative molecular studies, which offered "the greatest potential in comprehending the origin of eukaryotic cells and organelles."[12]

Raff and Mahler emphasized two features which they believed made the symbiotic theory of mitochondria awkward. First, according to the symbiotic interpretation, the protoeukaryote possessed many advanced cellular adaptations, yet it was supposed to be metabolically primitive, possessing inefficient aerobic energy-yielding pathways. They contested this, asserting that there were good indications that the protoeukaryote was already an aerobic respirer and would therefore receive no selective advantage from the incorporation of another organism (mitochondria/bacteria) with an aerobic respiratory system.

Second, geneticists had shown that cytoplasmic organelles such as mitochondria were not the whole organisms that some had previously conceived them to be. They were well integrated into the genetic systems of all organisms. Mitochondria lacked sufficient DNA to code for all of their own proteins; most were coded for in the nucleus. The "definitive proof" of their symbiotic nature or origin, their culturing outside the cell, was an experimental ideal that could never be realized. For its advocates, the ships had been placed into their bottles; for its critics, the bottleneck was the difficulty of believing that new genes could be incorporated into the nuclear genome through the inheritance of acquired bacteria. That is, if one accepted the SET, then there must have been a wholesale transfer of genes from the mitochondria to the nucleus—something Raff and Mahler found "difficult to conceive."[13]

These two criticisms attracted quick responses. The Nobel Prize–winning biochemist Christian de Duve, of the Université de Louvain and the Rockefeller University, entered the debate in 1973 to argue that the protomitochondrion which had entered the protoeukaryote had simply possessed a more efficient respiratory system than the host bacterium. Therefore its symbiotic acquisition would be advantageous even in an aerobic cell.[14] The next year, the microbiologist F. J. R. (Max) Taylor, at the University of British Columbia, addressed Raff and Mahler. His reply to their first criticism was essentially the same as that offered by de Duve. He pointed to recent advances in biotechnology to refute their second criticism: the transfer of bacterial genes to the nucleus of a higher plant was one of the most exciting areas of modern genetics.[15] Nonetheless, both issues remained at the center of debates about the SET. In 1977 Carl Woese suggested that the symbionts were originally photosynthetic nonsulfur bacteria and that their significance in terms of respiration arose later. But no example of endosymbiotic bacteria had been found which could serve as a model for this view.[16]

The argument for simplicity continued. In a subsequent overview, Taylor argued that before the 1960s autogenous hypotheses could well claim to be simpler and, a priori, more probable, because prior to the clear establishment of the major distinctions between prokaryotes and eukaryotes there were few structural or metabolic differences to explain. But as more fundamental differences became evident, there were more transitions that required explanation. Mitochondria and chloroplasts not only had been found to possess their own distinctive DNA, but also showed a closer resemblance to prokaryotes, and they behaved analogously to semiautonomous endosymbionts, reproducing themselves and being inherited in a non-Mendelian fashion. As a consequence, he argued, "autogenous views, or rather, what little had been detailed, began to appear more awkward."[17]

Taylor stated that autogenous hypotheses could not explain some *crucial* facts. DNA from plastids hybridized much more with rRNA of modern free-living cyanobacteria than with comparable surrounding cytoplasmic rRNA. In other words, plastids seemed to be more closely related to free-living protokaryotes than they were to the genetic entity in which they were embedded. This, he argued, was the reverse of what would be expected if plastids had become differentiated from cytoplasm, long after the divergence from cyanobacteria. For Taylor, such hybridization studies provided direct evidence of a symbiotic origin for chloroplasts.[18]

Yet, as Taylor well knew, these facts did not speak for themselves in favor of the SET. Their interpretation required certain assumptions. The key one was that base substitutions had occurred at an equal rate in all nucleic acids, irrespective of their location.[19] If organelle genomes had evolved at a slower rate than nuclear genomes, one could easily explain the similarity of organelles and prokaryotes without recourse to symbiosis. Moreover, it was certainly conceivable and indeed had been argued that cytoplasmic DNAs were not as readily subjected to evolutionary change as was the nuclear genome. It was possible that the different mode of reproduction of the nuclear genome, particularly the increased possibilities for recombination within nuclear genes and the assortment of new gene combinations made possible by sexual reproduction, had given rise to a more rapid rate of evolution of systems controlled by nuclear DNA than that of systems controlled by cytoplasmic organelle DNA.

Together They Stand; Divided They Fall

The endosymbiosis theory for chloroplasts and/or mitochondria could not rest on the strength of any single fact. All of the individual facts could be given alternative explanations. Some argued that the evidence better supported the endosymbiosis theory when mitochondrial and chloroplast origins were considered together. In 1975, Philip John and F. R. Whatley, at the Botany School, Oxford, brought forward evidence that the bacterium *Paracoccus denitrificans* resembled a mitochondrion more closely than did other bacteria; and they offered a "feasible" evolutionary transition from such an ancestral bacterium to mitochondria. Nonetheless, they still found that they could interpret their evidence in terms of the autogenous view if one considered the features common to bacteria and mitochondria to be "retained primitive states." They favored the endosymbiotic theory, asserting that it was "unrealistic to discuss the origin of mitochondria independently from that of chloroplasts" and that the endosymbiotic theory had "a greater applicability to explain the evolutionary origin of the eukaryotic cell as a whole."[20]

While the symbiotic interpretation of chloroplasts and mitochondria may have fared better when they were considered together, the same did not hold true for centrioles or basal bodies. Many biologists, while accepting the plausibility of the symbiotic origin of mitochondria and chloroplasts, doubted that the organelles concerned with mitosis and motility arose through symbiosis. This was true even for those, including Taylor, who defended the serial endosymbiotic hypothesis. The existence of DNA within basal bodies and centrioles was questionable, and detailed cytological studies revealed that they did not divide as did mitochondria and chloroplasts: new centrioles or basal bodies assembled from the general vicinity of existing ones; in some organisms they could be formed de novo. Moreover, during the late 1960s, electron microscopic studies of Jeremy Pickett-Heaps challenged the conventional view that centrioles were directly involved in mitosis; they seemed to be "along solely for the ride." Centrioles were not involved in the organization of microtubular systems; rather, they themselves were the product of primordial "microtubular organizing centers."[21] As Taylor remarked, "Pickett-Heaps's interpretation of MTOCs, which is an attractive one to this author, renders the exogenous origin of centrioles or other 9 + 2 homologs doubtful."[22]

There was also the problem of accounting for the origin of the nucleus. Perhaps the nucleus as a whole evolved as a symbiont from an ingested organism that subsequently lost much of its cytoplasm. Pickett-Heaps himself entertained this possibility in 1974: "Lest we consider such a suggestion far fetched, we should remember that even now, some dinoflagellates have two nuclei, one characteristic of dinoflagellates and the other a more typical eukaryotic nucleus."[23] As late as 1975, Tom Cavalier-Smith was still reluctant to accept a symbiotic origin for organelles, because it postulated too many separate events. He was especially critical of Margulis's theory of the orgin of cilia and the motitic spindle. In his view, her theory simply made "no structural and functional sense," and he dismissed as "incredible" her idea that the complexities of sex and meiosis evolved independently twenty-seven times.[24] He also opposed Margulis's view that the symbiotic origin of mitochondria was prequisite for the origin of the eukaryotic cell. He insisted that phagocytosis, which makes it easy for cells to acquire cellular endosymbionts, was limited to eukaryotes, and if this were so, then how could a hypothetical primitive amoeboid prokaryote play host to other primitive cells, as Margulis's theory proposed? Cavalier-Smith accepted the symbiotic theory only after proposing various important modifications to the details of the symbiotic theory and recognizing the existence of a new kingdom, the "Archezoa."[25]

In the second edition of their well-known text *The Plastids* (1978), Kirk and Tilney-Basset assessed the merits of the alternative theories for plastid origins in terms of simplicity. They relied on the now familiar argument that the more rapid rate of evolution of systems controlled by the nucleus could account for why the protein synthesis machinery of chloroplasts is so much more similar to the protein system of prokaryotes than to the protein synthesis system entirely under nuclear control.[26] They also insisted that the fact that many organellar proteins were under the control of nuclear DNA weakened the case for the symbiotic origin of plastids.[27] They concluded in despair:

> That the chloroplast is related, in an evolutionary sense, to the bluegreen algae seems very likely indeed. But whether it originated by encapsulation of

> the photosynthetic machinery within an ancient bluegreen algal cell, or whether it began as a bluegreen algal symbiont in an early eukaryotic cell, is impossible to say. Indeed, it is not obvious what kind of research will answer these questions.[28]

A Revival of Special Creation?

There was still one more tack to the argument about simplicity, illustrated in a rhetorical ploy made in 1974 by Thomas Uzell and Christina Spolsky in a paper entitled "Mitochondria and Plastids as Endosymbionts: A Revival of Special Creation?" Uzell, associate curator of herpetology at the Academy of Natural Sciences in Philadelphia, and Spolsky, a postdoctoral fellow of the American Cancer Society at the University of Pennsylvania, rebuffed remarks in the literature in favor of the endosymbiotic theory, stating that "the origin of eukaryotic organelles by endosymbiosis . . . is certainly no simpler than their origin by evolutionary divergence from prokarytiotic ancestors."[29] On the other hand, they argued on evolutionary grounds that the endosymbiotic theory was simplistic.

The central stake in the debate, as they saw it, was continuous versus discontinuous evolution, that is, whether organelles of eukaryotes arose not by "a gradual and continuous evolutionary process but rather abruptly, as a result of endosymbiosis."[30] By identifying evolutionary process solely with gradual change (neo-Darwinism) they constructed a conflict between "evolutionary rather than endosymbiotic origin."[31] They then claimed that the endosymbiotic theory of the origin of cell organelles was opposed to evolution in the same way special creation was opposed to evolution:

> The endosymbiosis hypothesis is retrogressive in the sense that it avoids the difficult thought necessary to understand how mitochondria and chloroplasts have evolved as a result of small evolutionary steps. Darwin's *On the Origin of Species* first provided a convincing evolutionary viewpoint to contrast with the special-creation position. The general principle that organs of great perfection, such as the eye, can evolve provided that each small intermediate step benefits the organism in which it occurs seems appropriate for the origin of cell organelles as well.[32]

At first glance one might suggest that Uzell and Spolsky's discursive strategy to dismiss the endosymbiotic theory of its scientific validity was based simply on emotive tactics—allying symbiosis with creationism and appealing to the hard-won legitimacy of evolutionary theory and the threat of anti-intellectualism and creationism which was emerging once again into prominence in the United States. The accusation that the endosymbiotic theory "avoid[s] the difficult thought" was certainly not based on the notion of simplicity as most participants in the debate understood it. In previous discussions, simplicity referred to the uniformity of theory, the fewest number of ad hoc explanations, and those that relied on the fewest assumptions when explaining an agreed set of facts. The central issue here is the agreed set of facts. What were to count as relevant facts, what phenomena were relevant to the debate, had preceded discussions about which account best explains the facts. Discussions of the simplicity of competing theories were limited to disciplines concerned with the study of cell organization.

However, when one enlarged the context to take into account all evolutionary phenomena, the endosymbiotic theory appeared to be ad hoc in the sense that it seemed to be invoked to account for a special case, the eukaryotic cell, only. The remarks of Uzzel and Spolsky highlight the insulated nature of the debates over the symbiotic theory of the cell. They underscore the extent to which cell and molecular biologists and evolutionary biologists of the 1960s and 1970s had overlooked the literature pertaining to symbiosis as a source of evolutionary change, the discussions about the generality of the phenomenon, and the debates about its underlying dynamics.

Are Origin Stories Scientific?

In the renewed debates over the symbiotic origin of organelles, many scientists had come to recognize that attempts to reach consensus on the basis of simplicity had failed. Yet there were still other methodological objections to claims that cell organelles evolved through symbiosis. From the outset of the debates of the 1960s and 1970s some had argued that theories of cell origins could never be tested and did not belong to the realm of science at all. The endosymbiotic theory was a historical theory. It attempted to account for the course of events that occurred more than a billion years ago in terms of present observations. There was simply no sure way of knowing whether what actually happened then was what is occurring now, and therefore one could deny for all such accounts any degree of scientific validity. The argument underlying this objection involves the whole question of historical science. It applies to any origin account, whether it refers to the origin of the eukaryotic cell, the origin of life, or the origin of the universe as a whole. To understand the arguments in this debate one has to take a closer look at the assumptions underlying scientists' approach to scientific theory and observations.

Ideally, scientific theory-making is based on the premise that if we know what nature is like at any one time, these laws tell how it will look at any *later* time. One makes observations, constructs theories, and tests the predictability of theories in terms of future experiments. Thus prediction is an attempt to correlate the present and future; that is, if *A* is observed, *B* will follow. Then there is the problem of origins. On the basis of present observations and theories tested against further observations, one attempts to reconstruct the past. However, such reasoning supposed that the "contingent causes" of a particular sequences of events were unchanging, were always operating in the same direction. In 1967, W. H. Woolhouse used this argument to nip accounts of the origin of cytoplasmic organelles in the bud, suggesting that that "the time has come to bury this kind of speculation with, by way of an epitaph, a parody of Wittgenstein's well-known remark, 'Whereof one cannot know, thereof one should not speak.' "[33]

Others, such as the microbiologist Roger Stanier, took a less severe attitude to speculations about organelle origins: though they belonged to the realm of metascience, they were relatively harmless, just so long as they did not become an obsession. "The most appropriate response to such speculations," he wrote, "(if they are plausible and logically consistent) is an Italian rejoinder, of which the amiable cynicism cannot be adequately translated: *Se non è vero, è ben trovato.*"[34] Although Stanier favored the endosymbiotic origin of mitochondria and chloroplasts, in 1970 he suggested that one could actually reverse the sequence of events and reason that prokaryotes originated from eukaryotic organelles:

> Is the comparative structural simplicity of prokaryotic organisms really indicative of great evolutionary antiquity? In view of their similarities to mitochondria and chloroplasts, it could be argued that they are relatively late products of cellular evolution, which arose through the occasional escape from eukaryotes of organelles which had acquired sufficient autonomy to face life on their own. This is a far-fetched assumption; but I do not think one can afford to dismiss it out of hand.[35]

Despite formal methodological arguments against them, evolutionary accounts—particularly the symbiotic theory—held great scientific appeal. And as Edward Minchin had argued in 1915, there were other criteria besides an unknowable ultimate truth for assessing the value of such theories. Theodosius Dobzhansky, often recognized as the "Darwin of the twentieth century," was frequently quoted for his assertion of 1973: "Nothing in biology makes sense except in light of evolution."[36] The symbiotic theory of the eukaryotic cell represented a unified theory which, all at once, embraced work done in a number of disparate disciplines, including microbial and cell biology, biochemistry, genetics, geology, marine biology, and paleontology, integrating them within a common framework of evolution.

Symbiosis research had always represented an integrative approach to life, which flew in the face of increased specialization and disciplinary boundaries in the life sciences. But during the 1960s many biologists had had enough of the increased specialization that had divided the life sciences. Thus in their book *Biogenesis of Mitochondria* (1968), Roodyn and Wilkie argued that at a time when biologists had become "so specialized and distant from one another," one should be appreciative of an "overall view" which the endosymbiosis theory, although admittedly speculative, encouraged.[37] Even Uzzel and Spolsky, who were so critical of the endosymbiotic theory of cells, credited it with renewing interest in the phylogeny of microorganisms: "Microorganisms cannot be fitted into the familiar categories of animals and plants, and attempts to group organisms into a more meaningful set of three to five kingdoms (Whittaker 1969) can only be heartily applauded."[38]

Supporters of the endosymbiotic theory looked for other criteria, such as fertility and plausibility, as alternative, or at least supplementary principles to argue for the value of the theory.[39] The merits of historical arguments were argued by Margulis in 1975. "The purpose of a scientific theory," she wrote, "is to unite apparently disparate observations into a coherent set of generalizations with predictive power." Margulis agreed with Woolhouse that "historical theories, which necessarily treat complex irreversible events, can never be directly tested." She argued that evolutionary biologists were in the same logical predicament as historians and "can only present arguments based on the assumption that of all the plausible historical sequences one is more likely to be a correct description of the past events than another."[40]

Though there were intrinsic limitations in proving any historical argument, one could test the relative strengths of one origin account against another with respect to their ability to unite apparently disparate phenomena. In this regard, Margulis maintained that the symbiotic theory provided an explanation of such phenomena as "the genetic behavior of organelles, physiology of the mitotic process, phylogeny of protists, gaps in the fossil record and in living biota, and even allows interpretation of certain atmospheric data."[41] Historical theory, and the serial endosymbiotic theory in particular, also rested its scientific validity upon its usefulness in generating new hypotheses which in turn led to creative scientific investigations. One could evaluate the competing theories based on this criterion as well. Even if the serial endosymbiosis theory should "eventually be proved wrong," Margulis argued, "it has the real advantage of generating a large number of unique experimentally verifiable hypotheses."[42] Margulis maintained these tenets as the basis of her classic text, *Symbiosis in Cell Evolution* (1981).[43]

Closing the Controversy

Despite all the discussion about formal methodological requirements of theory, the endosymbiosis hypothesis remained in doubt; to some, it seemed unlikely that its acceptance would ever become truly universal. It was still possible that the similarity of mitochondria and plastid traits to bacteria could be accounted for by arguing that plastid and mitochondria genomes had simply changed more slowly and less radically than nuclear genomes since their common divergence from a single ancestor in a single bacteria-like lineage. The fact that many organelle traits were encoded in the nucleus also weakened the endosymbiont theory. As the Harvard biologist Laurence Bogorad wrote in *Science* in 1975: "Recently acquired information about relationships between the organelle and the nucleo-cytoplasmic system forces us to consider more seriously than before the possibility that eukaryotism arose not by endosymbiosis but by another route."[44]

However, during the following decade new techniques based on nucleotide sequence analysis were developed; for many cell biologists, these techniques provided the rigor and closed the main controversy. The principal work was done by two groups: one headed by Carl Woese at the University of Illinois (Champaign-Urbana), and the other headed by W. Ford Doolittle and Michael Gray at Dalhousie University (Halifax, Nova Scotia). In the 1970s Woese and his group pioneered the use of ribosomal RNA sequencing analysis for phylogenetic studies and revolutionized the classification of bacteria. Based on ribosomal RNA sequences they distinguished two fundamental domains, the Archaebacteria and the Eubacteria.[45] Archaebacteria are for the most part restricted to peculiar ecological niches: anaerobic sludges for methanogens, acid hot springs for thermoacidophiles, and salt flats for halophiles. Nonetheless, since a few eubacteria occupy the same niches, molecular characteristics are the only reliable basis for distinguishing between the two groups.

As applied to the origin of organelles, the argument underlying molecular evolutionary analysis was this: if the nuclear genome and one of the cytoplasmic genomes could be shown to derive from genomic lineages which were phylogenetically distinct before the formation of the eukaryotic cell, then the symbiotic origin of that organelle could be taken as proven. For example, if plastid genomes comprise a branch or branches of the eubacteria—specifically, the cyanobacterial or oxygenic-photosynthetic-prokaryotic evolutionary tree—and if nuclear genomes arose from within the archaebacteria, then the xenogenous origin of plastids could not be questioned.

In the early 1970s, Linda Bonen, who had worked in Woese's lab, brought the technological expertise to Dalhousie University, where she worked with Doolittle. Bonen and Doolittle applied the technology first to cyanobacteria and then to the analysis of red algal chloroplasts.[46] Woese and his group did the same for *Euglena* chloroplasts.[47] Both research groups traced the prokaryotic nature of chloroplast ribosomal RNA to cyanobacteria. Subsequently, Bonen and Doolittle, in collaboration with Michael Gray and a graduate student, Scott Cunningham, applied the technique to cataloging ribosomal RNA of wheat mitochondria and provided further evidence of their prokaryotic nature.[48] Complete sequence analysis of wheat mitochondrial 18s ribosomal gene was later carried out in Gray's laboratory.[49] Woese's group subsequently used this sequence when they traced the origin of mitochondria to a specific group of prokaryotes, the alpha-proteobacteria.[50]

A nonsymbiotic origin for mitochondria and chloroplasts was less and less plausible. To maintain an endogenous origin of nuclear and organelle genomes from a common eubacterial ancestor, one would have to argue that this bacterial progenitor was *either* a cyanobacterium (in which case the mitochondria genome must have had an exogenous origin) or an alpha-proteobacterium (in which case the plastid genome must have originated oxogenously). That is, the progenitor genome in an endogenous origin scenario could not have given rise to *both* mitochondrial and plastid genomes at the same time. This line of argumentation had been laid out by Doolittle in 1980, in an article boldly entitled "Revolutionary Concepts in Cell Biology."[51] It was reiterated in 1982, in Gray and Doolittle's celebrated paper "Has the Endosymbiont Hypothesis Been Proven?"[52] However, at that time the question of mitochondrial origin(s) was still considered to be somewhat debatable, and the origin of the nuclear genome remained unclear. But ten years later Gray could boast that both of these points had been clarified. The alpha-proteobacterial ancestry of mitochondria was "certain" and there was increasing evidence pointing to an origin of the nuclear genome from a common eukaryotic-archaebacterial ancestor. Thus, in his overview of 1992, Gray confidently pointed to three distinct evolutionary lineages as the source of three genomes in the eukaryotic cell, ruling out any endogenous origin scenario.[53]

The basic problem, exogenous versus endogenous origin, is now considered to be rigorously and definitively resolved. But that did not exhaust the evolutionary questions and debates generated by the application of nucleotide sequence analysis to cyoplasmic genomes: How many independent symbiotic events gave rise to the diversity of modern-day plastids and mitochondria? Have plastids in some organisms evolved from eukaryote rather than prokaryote endosymbionts? Could one prove that genetic information brought into the cell from bacterial symbionts was transferred to the nuclear genome? Have organelle genomes evolved at radically different rates in the different major groups of eukaryotes?[54] These questions remained at the heart of a new burst of molecular evolutionary studies on cytoplasmic organelle genomes carried out principally in Canada beginning in the 1980s. In 1989 startling evidence presented by John Hall, Zenta Ramanis, and David Luck of the Rockefeller University indicated the presence of basal body/centriolar DNA, raising the possibility of rigorously testing Margulis's proposed spirochete origin of centrioles. However, evidence questioning the existence of basal body DNA was published the next year, which, as Gray saw it, "dimmed considerably" the prospect.[55]

Yet there was much more to studies of symbiosis than the origin of cell organelles. During the 1970s and 1980s, all those evolutionary and ecological questions that had been raised over the previous hundred years came to the fore in discussions of symbiosis. Were there general evolutionary mechanisms operating in nature which encouraged symbiosis? How should one understand the relations making up symbiotic associations? How should one think about the relations between species, between individuals, between organisms and cells, and between cells and genes? If symbiosis played a crucial role in the transition from prokaryote to eukaryote some two billion years ago,[56] did it play a major role in other transitions also, or was it a rare, exceptional, accidental phenomenon?

13

Is Nature Motherly?

> We consider naive the early Darwinian view of "nature red in tooth and claw." Now we see ourselves as products of cellular interaction. The eukaryotic cell is built up from other cells; it is a community of interacting microbes.
>
> LYNN MARGULIS, 1987[1]

The 1970s and 1980s were heady decades for evolutionary biology. Creationist movements emerged in many countries, dismissing evolutionary theory in a wholesale way.[2] Niles Eldredge and Stephen Jay Gould launched their theory of "punctuated equilibria" arguing against gradualism, that evolution was interspersed with periods of rapid speciation, perhaps resulting from "hopeful monsters."[3] A new discipline of sociobiology was born, proclaiming that it could account for cooperation and much of human social evolution in terms of gene selection, and asserting that we are born selfish. With environmentalists warning that through our increased industrial activities we are acting out a great tragedy that may lead to our destruction, and with the threat of nuclear war continuing, "extinction studies" emerged as a field in its own right. Catastrophe theorists argued that "a great clock in the sky is controlling our biological destiny." They warned that a death star, named Nemesis after the ancient Greek goddess of retribution and justice, had caused the extinction of dinosaurs and will return in a few million years.[4] Still others protested against what they saw as the Newtonian mechanical worldview underlying neo-Darwinian theory, which separated organisms from their environment and treated them as if they were mechanical objects acted on by selective forces in the environment. Referring to the goddess of mother earth, Gaia, they argued that the evolution of organisms and their physical environment were tightly coupled into a global adaptive control system. The scope and significance of symbiosis as a source of evolutionary innovation became associated with many of these issues.

The Human Nature of Origin Stories

During the 1970s, while cell biologists debated the criteria for proof for the endosymbiotic theory, others had already declared a revolution and had begun to discuss its larger evolutionary meanings. Formal methodological arguments designed to dismiss evolutionary speculation in a wholesale way missed an important aspect of the general nature of origin stories. They were not simply descriptive accounts of what happened millions and millions of years ago. They were also prescriptive. The appeal to symbiosis as a natural law upon which to base our relations to each other and to other species continued throughout the 1970s and 1980s. This is not to say that the message of symbiosis became any clearer or that there was any general agreement about the principles underlying the phenomenon. That mitochondria and chloroplasts originated as symbiotic bacteria was recognized by many biologists as a revolution in scientific thought. But it did not entail a revolution in biologists' conceptions of the laws of nature. Interpretations of the underlying dynamics of symbiosis, its scope, and its significance continued to be laden with charges of seemingly inescapable anthropomorphism and politics.

In 1970, the well-known biochemist Seymour Cohen suggested that governments might be interested in the "revolution" presented by the endosymbiotic theory of cells:

> Just as the Copernican revolution demonstrated that man is not the center of the Universe, so the investigation of this problem may show that a man (and indeed any higher organism) is merely a social entity, combining within his cell the shared genetic equipment and cooperative metabolic systems of several evolutionary paths. We suspect that governments should be interested in such a possibility, although their responses many not be readily predictable.[5]

Indeed, one could not predict what they would make of it. In his well-known text *An Introduction to Molecular Genetics* (1971), the phage geneticist Gunther Stent expressed his view of the politics of the cell in terms not of international cooperation for mutual benefit but of its exact "social Darwinist" antithesis:

> Thus a eukaryotic cell may be thought of as an empire directed by a republic of sovereign chromosomes in the nucleus. The chromosomes preside over the outlying cytoplasm in which formerly independent but now subject and degenerate prokaryotes carry out a variety of specialized service functions.[6]

The master–slave interpretation of the relationship was the common view of biologists, according to Lewis Thomas. In his award-winning book *The Lives of a Cell* (1974), Thomas discussed the disquieting nature of the quiet revolution that claimed organelles "were strangers living within us," that we did not simply descend from bacteria but are constituted by them. But as he saw it, their conceptual construction as "enslaved creatures," captured to provide ATP for cells, or to provide carbohydrate and oxygen for cells equipped for photosynthesis, was merely a one-sided subjective anthropocentric way of looking at them. "From their own standpoint," he wrote, "the organelles might be viewed as having learned early how to have the best of possible worlds, with least effort and risk to themselves and their progeny. . . . To accomplish this, and to assure themselves the longest possible run, they got themselves inside all the rest of us."[7] Far from being a case of one-sided exploitation, Thomas assured his readers,

> there is something intrinsically good natured about all symbiotic relations, necessarily, but this one, which is probably the most ancient and most firmly established of all, seems especially equable. There is nothing resembling predation, and no pretense of an adversary stance on either side. If you were looking for something like natural law to take the place of the "social Darwinism" of a century ago, you would have a hard time drawing lessons from the sense of life alluded to by chloroplasts and mitochondria, but there it is.[8]

Our fear of microbes as enemies of life was equally misfounded, in Thomas's view: it was a paranoid delusion on a societal scale, reinforced and maintained by the interests of chemical companies that encourage us to believe that we stay alive and whole through diligence and through disinfectants designed to kill germs and plastics designed to seal them off. It was also due, in his view, to "our need for enemies" and to our memory of what things used to be like. Thomas emphasized how, only a few decades earlier, bacteria were a genuine household threat and all of us were always aware of the nearness of death. We had lobar pneuomonia, meningococcal meningitis, streptococcal infections, diphtheria, endocarditis, enteric fevers, various septicemias, syphilis, and always, everywhere, tuberculosis. "Most of these have now left most of us, thanks to antibiotics, plumbing, civilization, and money, but we remember."[9]

But pathogenicity had never been the rule. As Thomas saw it, microbes were not coming after us for profit; symbiosis was not a matter of disease gotten under control, rather disease resulted from failed symbiosis—"an overstepping of the line by one side or the other, a biological misinterpretation of borders."[10] He emphasized that staphylococci live all over us; that most bacteria were grazing, altering the structure of organic molecules so that they become usable for the energy needs of other life forms. In Thomas's estimation, we had more to fear from our own immune system.[11]

Back to Gaia

It wasn't just medical triumphs that promised to lessen our fear of microbes during the 1970s. Their proposed transformation from enemies to allies also resulted from extending evolutionary studies beyond the past 600 million years of plants and animal evolution to the preceding 3000 million years of microbial evolution.[12] The view of chloroplasts and mitochondria as organisms collaborating for the common good, that life had not only descended from bacteria but was constructed by them, combined with NASA-based studies of the mid-1960s on whether there was life on Mars to bring about an increased awareness of the work carried out by microbes in creating the atmospheric conditions for existing forms of life. This was reinforced in the early 1970s with theoretical discussions instigated by NASA to consider the possibility of making the planet Mars habitable for human life.[13]

Visions of the Earth from the distance of space, the production of the atmosphere by the biosphere, and visions of continents moving from the distance of geological time inspired, in some, an integrated and holistic appreciation of life. "The earth," Thomas wrote in 1974, has the "organized, self-contained look of a live creature, full of information, marvelously skilled in handling the sun."[14] The atmosphere which bound the earth, he asserted, constituted the "world's largest membrane."[15]

The British scientist and inventor James Lovelock had a similar revelation of life on Earth as one complex adaptive system, which he called Gaia. Lovelock's concept of Gaia began to take shape in the mid-1960s when he was working for NASA in the Jet Propulsion Laboratory just outside Pasadena, California.[16] The quest was to discover whether there was life on Mars. Instead of a local search at the site of landing, he proposed to analyze the chemical composition of the planet's atmosphere. If Mars were lifeless, then it would have an atmosphere determined solely by geochemical phenomena and would be close to the chemical equilibrium state. But if it bore life, he and the philosopher Dian Hitchcock reasoned, the atmosphere would be in a state of chemical disequilibrium since organisms at the surface would use it as a source of raw materials and as a depository for wastes. Their examination of the atmospheric evidence from the infrared astronomy of Mars showed that its atmosphere was close to chemical equilibrium (composed of 98 percent carbon dioxide, 2 percent nitrogen, with only traces of oxygen) whereas the Earth was in a state of deep chemical disequilibrium (with only traces of carbon dioxide, 79 percent nitrogen, and 21 percent oxygen). For Lovelock, the lifelessness of Mars only seemed to make Earth appear more alive.

Gaia, a view of Earth from space, was a spinoff of space research. But according to the Gaian hypothesis, the abiotic conditions for life were not simply a spinoff of the biosphere. On the contrary, it supposed that the atmosphere, the oceans, the climate, and the crust of Earth are regulated at a state comfortable for life by and for life itself. In the three or four billion years since life began, the chemical composition of the Earth's atmosphere and oceans never altered radically enough to destroy all life. This demanded explanation; for Lovelock, it required the existence of an active control system. He first formally put forward the idea of Gaia as a control system in 1972. Soon he began to work with Lynn Margulis and in 1973 they defined the Gaia hypothesis as "the notion of the biosphere as an adaptive control system that can maintain the Earth in homeostasis."[17]

When Lovelock introduced Gaia in his book of 1979, he contrasted this concept to what he considered to be the conventional geochemical wisdom which held that life "adapted to the planetary conditions as it and they evolved their separate ways."[18] Like Spencer's superorganism and Reinheimer's Gaia, Lovelock's top-down perspective focused on how the components of Gaia function for the good of life as a whole.[19] Like other cooperative models of nature before it, Lovelock's Gaia offered an optimistic view of nature and the place of humans in it. It was presented as a nonanthropocentric alternative to "that pessimistic view which sees nature as a primitive force to be subdued and conquered" and to the "depressing picture of our planet as a demented spaceship, forever travelling, driverless and purposeless, around an inner circle of the sun."[20]

Lovelock explored the relevance of the Gaian hypothesis to the effects of human interventions into nature. Rachel Carson, the fountainhead of the contemporary environmental movement, had marshaled substantial scientific evidence of the threat to life posed by the excessive use of pesticides and protested against those scientists who promised ever-larger bombs and those scientists who continued to pride themselves in a vision of dominating and controlling nature. Carson's revolutionary book, *Silent Spring* (1962), inaugurated the literature on ecological apocalypse.[21] But Lovelock opposed the views of those environmentalists who believed that human activity was leading to the destruction of the whole world, and that our only means to escape this great tragedy was to renounce technology, and especially nuclear energy.[22]

Environmentalists often had their priorities and concerns misplaced, Lovelock argued, because Gaia's cybernetic system of regulation made her more resistant to pollution than commonly believed. He argued that the accumulation of DDT throughout the Earth was not as great as was expected and that there were natural processes for its removal which had not been anticipated when the investigations began.[23] In his view, one had little to fear from chlorofluorocarbons, which had been "unjustifiably" banned in the United States, since "even on the gloomiest prediction, ozone depletion is a slow process."[24] He insisted that far from threatening the environment, science and technology were necessary to define and overcome the dangers. After all, he remarked, Rachel Carson had been heard; one now takes heed of the overuse of pesticides.

Lovelock did worry about the possibility of destroying Gaia, but not in terms of nuclear power or the sloppy use of spray cans, which he considered to be less important than the threat from extinction of species caused by human overpopulation. He warned that it is in the less-watched tropical rainforests and the seas close to the continental shores where harmful practices may be pursued to the point of no return. In the long run, he argued, one had to guard against the possibility that Carson was right for the wrong reasons. If a silent spring bereft of bird song may ever arrive, it will not be due to direct poisoning of birds by pesticides, but because of the overpopulation of humans which results from the use of such pesticides. As he bluntly wrote, "There is only one pollution . . . people."[25]

Dismissing what he took to be a reactionary "back to nature campaign,"[26] Lovelock supported the "hopeful, optimistic view and liberal views" expressed by René Dubos. In 1976, Dubos also argued for a "symbiosis between earth and humankind" as a creative alternative to preserving "pristine systems" at one extreme and exploiting the planet for short-term economic profit at the other:

> The goal of this relationship is not the maintenance of the status quo, but the emergence of new phenomena and new values. Millennia of experience show that by entering into a symbiotic relationship with nature, humankind can invent and generate futures not predictable from the deterministic order of things, and thus can engage in a continuous process of creation.[27]

Dubos had always disliked the expression "spaceship Earth." It called to mind "a mechanical structure carrying a limited supply of fuel for a defined trip and with no possibility of significant change in design." It only led to faulty thinking about environmental problems.[28] He agreed with Lovelock that "the earth has many attributes of a living, evolving organism. It constantly converts solar energy into innumerable organic products and increases in biological complexity as it travels through space."[29] But for Dubos, Lovelock's emphasis on homeostasis did not do full justice to the Gaia concept, the coevolutionary process by which organisms transform the surface of the Earth while themselves undergoing continuous change.

Dubos suggested that the Gaian control might be punctuated, resulting in global homeostasis only over a period of time which is short on the evolutionary time scale. After all, almost all the examples Lovelock discussed referred to creative evolution rather than homeostatic reactions. For example, the accumulation of oxygen in the air, which became significant two billion years ago as a result of photosynthetic activities, probably destroyed many forms of life for which this gas is poisonous. Furthermore, the process of change may pick up speed and complexity as a result of human interventions, and Dubos emphasized that on a local level, profound coevolutionary changes have occurred in certain terrestrial environments and in their biological systems during historical times. This was the key point Dubos featured in his book *The Wooing of the Earth* (1980).[30] While acknowledging the destruction humans have wrought, he emphasized how much of what we regard as "Nature" is human-made, the resiliency of the natural world and our own ability to enrich it and to ensure "the harmonious interplay of mankind and Earth."

Mother Nature, Antiscience and Myth

During the 1980s Gaia was discussed favorably in semipopular science journals such as *New Scientist* and *Scientific American* and conferences were held to discuss its political and philosophical implications.[31] By the end of the decade, Lovelock's Gaia hypothesis had attracted considerable attention beyond biological circles. Theologians and holistic philosophers showed an interest in Gaia, and in Lovelock's view, "Greens" also preferred it to reductionist biology.[32] But most mainstream biologists remained silent. Lovelock saw himself as an independent scientist working on the fringe of a defensive scientific community, intolerant of radical ideas. Yet, as he recognized, even his "friends among biologists" became anxious when he spoke of "Gaia as an organism." Indeed, Margulis herself opposed Lovelock's view that "Gaia is an organism." In her view, it only encouraged critics and cranks to flourish and alienated those scientists who should be working in a Gaian context.[33] Moreover, she argued, no one had defined "organism" in this context. Later she employed the term "autopoiesis" (from the Greek *auto,* self, and *poeisis,* to make) to refer to the self-maintaining activities of living things.[34] But in referring to Gaia as an organism, Lovelock believed he was continuing a tradition that was rejected in the nineteenth century when biology became ever more compartmentalized, "imprisoned in a narrow, almost puritan, reductionism, a consequence perhaps of the long and continuing war with vitalists, animists and religious fundamentalists."[35] These issues were apparent in many critiques of Gaia.

In 1988 the British biologist John Postgate "broke the silence" to complain about the attention Gaia received in the popular press. "Gaia—the Great Earth Mother! The planetary organism! Am I the only biologist to suffer a nasty twitch, a feeling of unreality, when the media invite me yet again to take her seriously?"[36] Postgate denied Lovelock's assertions about the long neglect of the Gaia metaphor and insisted that there was nothing in it—except the name—that had not been utterly familiar to biologists for all of this century. He contended that it was axiomatic in biology that biological systems modify the environment to suit themselves and that evolutionary changes in the living and nonliving components of the Earth were a necessary consequence of their interactions. That was why most biologists remained politely silent when Lovelock himself also grasped these ideas. But when the colleague proposed a "Goddess of Wheeliness," Postgate remarked, "We shake our heads sadly" and when "Wheely-Goddess worshippers start popping up all over, then it is definitely time to worry. A principle of wheeliness, like a planet sized organism, may be fun as imagery; as anything more scientific it is silly and dangerous."

It was dangerous, as Postgate saw it, because it was "psuedoscientific mythmaking" and it could only do further damage to the repute into which science had fallen over the previous two decades. He did not explain the complex reasons for antiscience attitudes, but the consequence of that disrepute was all too apparent: "Fringe science, pseudoscience, obscurantism, wishful thinking and mysticism today find almost medieval favor even among educated people." The interest in Gaia was part of a wave of alternatives to established scientific knowledge: "the surge of astrology, fringe medicine, faith healing, nutritional eccentricities, religious mysticism and a thousand other fads and cults which now plague developed societies." Postgate's attitude toward the organismic view of Gaia was representative of many biologists.

When Lovelock's first book appeared in 1979 many evolutionists rejected Gaia as a myth, not simply because it only encouraged antiscience movements, but because of their inability to conceive of any mechanism to account for the evolution of such a coordinated whole. The concept that the biosphere evolved as a whole was, of course, not new to twentieth-century biology with Lovelock's Gaia hypothesis. It had been maintained in such well-known reference texts as Allee et al.'s *Principles of Animal Ecology,* first published in 1949. As Alfred Emerson wrote in the chapter on "The Evolution of Interspecies Integration and Ecosystem":

> Through the action of the habitat upon living systems, the reaction of these upon the environment often resulting in an organic evolution of the physical environment . . . we find that life and habitat are integrated into an evolving ecosystem, ultimately incorporating the entire biosphere of the earth. The unity of the biosphere is the resultant of the complex interactions of many factors—a complexity so great that many competent biologists have failed to recognize the existence of the unitary whole.[37]

Richard Dawkins and W. Ford Doolittle made the most determined effort to show that evolution could not lead to cooperation on a global scale, as Gaia implied. In the *Extended Phenotype* (1982), Dawkins ridiculed "Gaia's webs of mutual dependence and symbiosis" as pop ecology: it was merely a moralistic exhortation with an unsound theoretical basis.[38] Nonetheless, as Dawkin's saw it, since the Gaia hypothesis had been enthusiastically espoused by "no less a scientist than Margulis" and praised by Dubos and a few other prominent biologists as a daring but tenable hypothesis, it could not be ignored.[39] Doolittle's attitude was similar. In 1981, in a review entitled "Is Nature Really Motherly?" he likened the theory to old-fashioned wishful thinking: "It gives one a warm comfortable feeling about Nature and man's place in it."[40] Doolittle and Dawkins did not doubt the existence of some of the feedback loops which Lovelock claimed to exist. Methane production may now balance oxygen production, but they rejected the idea that such balances were created by natural selection or that they were anything but accidental. It was all right to speak of cells of an individual organism as contributing to a common good. But the biosphere was not a self-replicating individual; it had no coherent heredity, no physical continuity through space and time.

The fatal flaw in Lovelock's model, as they saw it, was the lack of consideration of the level at which natural selection acted. This issue, prominent in critiques of the Gaia hypothesis emerged at the heart of many theoretical discussions of the origin and maintainance of classic cases of "mutualistic symbiosis." During the 1970s there had already been heated controversy among evolutionary biologists about the precise nature of the unit of selection.[41] The favored choices were genes within organisms; organisms within populations; populations in species. Does selection act at several levels of genealogical hierarchy including genes, organisms, and populations in species? How cooperation was to be understood was a central stake in this debate. Some maintained that entities at each level of the hierarchy can act as biological "individuals," and that selection operates at all three levels. According to this view, behavioral traits that were detrimental to the individual but beneficial to the species may be maintained within a population by natural selection operating on the population as a whole.

In the mid-1970s, however, sociobiology emerged as a new discipline led by the writings of E. O. Wilson, John Maynard Smith, William Hamilton, Robert Trivers, and Richard Dawkins. This new discipline attempted to account for cooperation and different aspects of social organization and behavior on the premise that evolution acts to maximize an individual's genetic representation in the next generation. Sociobiologists extended their hereditarian and evolutionary theories to account for important aspects of human social behavior as well. Ethics, aesthetics, politics, culture, and religion all fell within the scope of sociobiological inquiry.[42]

In *The Selfish Gene,* Dawkins argued that selection operated at the lowest level of the hierarchy: the gene. All behavior traits would be selected for, if the trade-off value resulted in more benefit than harm to organisms with similar genes. When altruistic behavior occurred between closely related individuals, usually between immediate family members, sociobiologists invoked "kin selection": although altruistic behavior may lead to individual death, it increases the probability of survival of genes the altruist shares with its relatives in the population. This could explain sterility in insect colonies and the suicidal barbed sting fo the honeybee worker. Similarly, if a human mother risked her own life to save her child, it was because she was contributing to the survival of her own genes "invested" in the child. Thus all altruistic behavior was really a form of selfishness.

Sociobiologists applied the same genetic cost-benefit analysis when altruistic behavior was exhibited between organisms in which relatedness was low, such as grooming among birds and mammals or when one human saves another from drowning. In these cases, they employed the theory of reciprocal altruism as developed by Robert Trivers in 1971. The theory of reciprocal altruism proposed that natural selection has operated such that risk-taking acts of kindness can be recognized and reciprocated so that the net fitness of both participants is increased. Thus all such individual altruistic acts are actually carried out with the expectation of a payoff to that individual later on. As Trivers put it, "Reciprocal altruism can also be viewed as a symbiosis, each partner helping the other while he helps himself. The symbiosis has a time lag however, one partner helps the other and must then wait a period of time before he is helped in return."[43] Trivers suggested that many of our psychological characteristics, such as envy, guilt, gratitude, and sympathy, were genetically determined and shaped by natural selection for improved ability to cheat, detect cheats, and avoid being thought to be a cheat.

As Dawkins saw it, "Money is a formal token of delayed reciprocal altruism."[44] And as T. H. Huxley had argued generations earlier, Dawkins maintained that there was a discontinuity between cosmic evolution and human morality. "Be warned that if you wish as I do, to build a society in which individuals cooperate generously and unselfishly towards a common good, you can expect little help from biological nature. Let us try to teach generousity and altruism, because we are born selfish."[45]

In *The Selfish Gene* Dawkins discussed symbiosis in a chapter entitled "You Scratch My Back, I'll Ride Yours." He viewed the idea that mitochondria originated as symbiotic bacteria as "one of those revolutionary ideas . . . whose time had come." However, Dawkins redefined symbiosis. The word had been conventionally used for associations between different species. But Dawkins reasoned that since there was no selection for "the good of the species," there seemed to be "no logical reason to distinguish associations between members of different species as things apart from associations between members of the same species."[46] All of Dawkins's arguments were dedicated to the same reductionist point: that cooperation could be explained as a winning strategy, through which an individual, "blindly programmed to preserve selfish genes," can best promote its own survival. As he put it, "In general, associations of mutual benefit will evolve if each partner can get more out than he puts in. This is true whether we are speaking of members of the same hyena pack, or of widely distinct creatures such as ants and aphids, or bees and flowers."[47]

It was this issue of cost-benefit accounting that Doolittle and Dawkins found to be the most serious problem for the evolution of Gaia, where the effects of an organism's actions would be measured only thousands and millions of years later. The organism which changed tthe atmospheric conditions did so for immediate or short-term benefit, not for the good of the community, not for generations to come. As Doolittle remarked: "It is certainly correct to say that no serious student of evolution would suggest that natural selection could favor the development in one species of a behavior pattern which is beneficial to another with which it does not interbreed, if this behavior were either detrimental or of no selective value to the species itself."[48] Dawkins made the same point in terms of human relations:

> Entities that pay the costs of furthering the well-being of the ecosystem as a whole will tend to reproduce themselves less successfully than rivals that exploit their public-spirited colleagues, and contribute nothing to the general welfare. Hardin (1968) summed the problem up in his memorable phrase "The tragedy of the commons" and more recently (Hardin 1978) in the aphorism, "Nice guys finish last."[49]

It was difficult enough to account for the stability of associations in which the effects might be immediately beneficial to all partners. Kinship theory predicted that virtually all cases of altruism and most cooperation (apart from its appearance in humans) would occur among closely related individuals, usually between family members.[50] Based on mathematical and other formal analyses, leading ecologists and theoretical evolutionary biologists continued to insist that mutualistic interactions between two or more species were unstable, and consequently rare or ecologically and evolutionarily unimportant.[51] Some ecologists considered them to be more prevalent in stable, nonseasonal, nonfluctuating environments, such as the tropics, than in temperate regions.[52]

Tit for Tat

That mutualism was really a form of selfishness in which individuals are motivated by an expectation of reciprocal reward was theoretically easy to imagine among organisms which were capable of recognizing and remembering each other as individuals. But it was unlikely that "lower organisms" were making such discriminations when they entered into such relations, and genes themselves had no foresight. After thinking about evolution of cooperation in human social relations, Robert Axelrod, a political scientist, teamed up with William Hamilton, a biologist, to elaborate its implications for biological evolution. To demonstrate how, without foresight by the participants, such "mutually advantageous symbiosis" as the lichen could originate and be maintained, they turned to game theory and computer models.

Game theory was developed and first applied by economists and military strategists. But in the 1970s and 1980s computer games became a major tool for those evolutionists who tried to account for how cooperation could emerge in a world of egoists without central authority. Solutions to social relations and strategies were sought by creating a model system of conditions and then programming a computer to work through the moves and the interactions until a statistically significant sample results. Axelrod and Hamilton thought that the game-theoretic approach could also account for the transformation, or rather "defection," of resident bacterial symbionts into pathogenic parasites, or latent viruses into cancer-causing agents, if a human host becomes old or seriously ill.[53]

Their conclusions about the evolution of cooperation were formed by pitting various behavioral strategies against one another in a round-robin computer tournament of the famous prisoner's dilemma game, based on two egoistic players reponding to one another's previous moves of defecting and cooperating in a predominantly noncooperative environment. For long-term evolutionarily stable relations, Tit-for-Tat, developed by Anatol Rapoport, was the simplest and best computer program. It begins with a cooperative choice and from then on does what the other player did on the previous move.

Tit-for-Tat relied on individuals' keen ability to recognize one another, and thus to be able to retaliate against a defecting individual. Axelrod and Hamilton reasoned that lower organisms, which did not possess this ability, had to drastically lower the number of individuals with which they interact in order for the game to apply to them. Microbes could accomplish this by maintaining continuous contact with one player. This would account for most "mutualisms" between members of different species, such as a hermit crab and its sea anemone partner, a cicada and the colonies of microorganisms living in its body, and a tree and its mycorrhizal fungi. Another way of avoiding the need for recognition was to have a fixed meeting place. For example, aquatic cleaner mutualisms, in which a small fish or crustacean removes or eats parasites from the body or mouth of a large fish predator, occur in coastal and reef situations where animals live in fixed home ranges or territories.[54]

Thus Hamilton and Axelrod concluded that "Darwin's emphasis on individual advantage has been formalized in terms of game theory."[55] However, Axelrod's book has a much broader range, with a chapter on "the live and let live system" in trench warfare in World War I. It also contained advice to national leaders about developing a more cooperative world: "Don't be envious, don't be the first to defect, reciprocate both cooperation and defection, and don't be too clever."[56] This was the watchword that came from computer games.

If such computer models could show how evolutionarily stable forms of cooperative interaction between two species could evolve without foresight of the participants, it was impossible on this basis to reason how a self-regulating biosphere could have evolved this way. In order for mutualistic symbiosis to arise, according to an interaction rule such as Tit-for-Tat, both partners had to receive benefit, and each evolutionary step taken by one or the other partner in the development of a symbiosis must be immediately and individually beneficial to that partner.[57] But the species comprising the partnership of Gaia were immense in number and diverse in kind, and the time scale over which the actions of each member affect the welfare of all exceeded by many orders of magnitude the generation of any. How could they collaborate as one whole, when, as Doolittle put it, "the consequences of defection by one species will not be perceived by others for generations or millennia"?[58]

To account for the error in Lovelock's reasoning, Doolittle employed the anthropic principle, an argument that began to be taken seriously by cosmologists in the 1970s.[59] Basically, it asserted that the possibility of observing life itself ruled out the possibility of observing other values than those which permitted life itself. In other words, evolution looked as if it were goal-seeking and life looked as if it were controlling the conditions of life for its own ends, because if it were not like that we would not be here to think about it. In Doolittle's words, "Only a world which behaved as *if* Gaia did exist is observable, because only such a world can produce observers."[60]

Delimiting Symbiosis

The Gaian hypothesis and theories on the evolution of cooperation have taken us far afield from experimental studies of symbiosis. But in many ways the issues to be discussed were similar: What counts as an individual, a unit of selection? What are the mechanisms for perpetuating symbiotic complexes? What is the role of symbiosis in evolutionary change, and how should symbiosis be defined? There was a great deal of theoretical discussion of mutualistic symbiosis, but what was needed, experimentalists argued in the 1970s and 1980s, were more actual descriptions of how organisms really interact, experiments with what actually happens when a potential mutualist is removed.[61] Still others emphasized that it may be difficult to demonstrate mutualistic symbiosis in general.

Despite assertions about mutualistic symbiosis and reciprocity theory by theorists and sociobiologists, David C. Smith and Angela Douglas asserted in their textbook, *The Biology of Symbiosis* (1987), there had been "no complete and rigorous experimental proof that any symbiosis is mutualistic."[62] That is, it was often straightforward to show that the host grows and reproduces faster in the presence of a microbial symbiont, and if this could be called "benefit," with the exception of root symbionts, there was still no evidence to show that the microbial partner benefited. One also had to consider that certain associations may be both parasitic and mutualistic at different stages or under different environmental conditions. Given these contingencies and that "benefit" itself may sometimes be difficult to define, many contemporary symbiosis researchers reinvoked a definition of symbiosis that did not rely on a rigid distinction between parasitism and mutualism.[63]

Despite theorists' attempts to reduce symbiosis to relations among individuals or their genes, symbiosis researchers continued to define symbiosis in terms of the relations between different species that are physically associated. The development of a newly defined field of experimental studies, and the establishment of a niche within contemporary evolutionary biology, was one of the stakes in this delimitation of the term. With all its conceptual hazards, specialization and the formation of new disciplines with their own aims, doctrines, theories, and techniques is an effective social strategy—as a locus for exchanging information, training new generations of researchers, and as a locus for funding. The lack of this social structure for studies of symbiosis was obvious. The coherence of the natural world advocated by some symbiosis theorists stood in dramatic contrast to the incoherence of symbiosis studies. Delimiting the field of symbiosis was no easy task. However, during the 1980s the boundaries of a new discipline were beginning to be drawn.

In 1983 an International Society of Endocytobiology (ISE) was founded by the German biologists Werner Schwemmler and H. E. A. Shenck, and with its own journal, *Endocytobiosis and Cell Research,* for studies in intracellular symbiosis. It was rooted in the conceptual consensus regarding the symbiotic origin of organelles.[64] During the 1980s the ISE published three international colloquia. In 1985, the journal *Symbiosis* was founded, edited chiefly by the Israeli lichenologist Margolith Galun, dealing with issues ranging from lichen biology and marine symbiosis to the potential industrial uses of symbiotic microorganisms. (Symbiosis researchers are presently in the midst of a merger of these two journals, to unite those scientists studying all kinds of symbiosis with those investigating intracellular organelles.)

By the end of the decade those who studied symbiosis could boast, with David Smith, that their subject had advanced "by leaps and bounds over the previous 30 years." Those who once confined themselves to the study of a single type of association such as coral reefs, insects, legumes, or lichens, he declared, now draw on ideas and concepts from other associations, giving the "whole subject much-needed cohesiveness."[65] Far from being an exceptional phenomenon, symbiosis researchers could point out that 90 percent of land plants are myccorrhizal in nature, and virtually all mammalian and insect herbivores would starve without their cellulose-digesting symbionts. The transmission of microbial symbionts from one generation to the next could be accomplished by a variety of means. In many cases, the acquired symbionts had become hereditary.

While experimental studies in symbiosis begin to take the form of a discipline, researchers continue to lament the relative obscurity of their field and the lack of funding for studies in symbiosis per se. This is not to say that there were no funds for studies involving symbiosis. For example, agricultural and forestry research funds have fostered analyses of nitrogen-fixing bacteria and legumes and mycorrhizae; oceanography and marine science programs have involved the study of symbiotic algae in corals. In the United States the Office for Naval Research has assisted in studies of bacteria habored in the light organs of luminous fish.[66] Yet symbiosis is not recognized as a field unto itself with any independent standing, and as Margulis has remarked, "mention of symbiosis in a grant application tends to deny funding."[67]

Margulis suspects that the reason for this is that symbiosis continues to be laden with political messages, such as Kropotkin's well-known mutual aid arguments. Furthermore, "the healthy, positive, perhaps even feminine connotations of symbiosis and mutualism have suggested that research on these topics is relatively unimportant."[68] But there was more than this behind the move to lead experimental studies of symbiosis (and the concept itself) away from an emphasis on cooperation. Lumping investigations of symbiosis together with studies of mutualism, or with forays into sociality, only served to obscure consideration that symbiosis was a means of generating variation in evolution. Sociobiologists who discussed symbiosis in terms of reciprocity theory and selfish genes failed to address the morphological and physiological changes brought about by intimate associations between species.

During the 1980s, symbiosis researchers extended symbiosis from accounting for the transition from prokaryote to eukaryote to explore its role in other transitions as well. They asserted that intimate symbiotic associations brought together genetic material from very distantly related organisms, and thus the changes arising from them are far greater in magnitude than those which result from the gradual accumulation of differences within species (gene mutation, recombination, etc.).[69] It was obvious that the integration of a symbiont may require reciprocal adjustments involving gene mutation and selection. But all these mechanisms had to be considered in any complete theory of evolution. The importance of the acquisition of additional genomes as a mechanism of evolution in bacteria had been widely discussed.[70] However, such horizontal gene transfer between prokaryotes and eukaryotes, and between eukaryotes of different species of eukaryotes, was generally considered by evolutionists to be a rare phenomenon.[71] Today, some researchers, such as Paul Nardon and Anne-Marie Grenier, consider the acquisition of a symbiont to be "the most important and powerful force in evolution."[72] Others continue to assert that the highly interactive interspecies community, the raw material for symbiotic associations, is not a general feature of nature. Thus Richard Law argues that while symbiotic phenotypes do come about from time to time with profound consequences for the course of evolution, they are still much less common as a source of evolutionary innovation than changes generated within species.[73]

Concluding Remarks

We have located studies of symbiosis peering through the cracks and creeping across the boundaries which separated ecology from evolution, plants from animals, health from disease, nurture from nature, and the individual from the community. In so doing, we have uncovered layers of oppositions, doctrines and disciplines, and diverse phenomena that have led to disparate interpretations of symbiosis, its scope and significance. In summarizing them here, we see that this history is not a matter of peeling off obstacles to come closer to some hidden core of naked truth. Symbiosis is as much like an onion today as it was a century ago.

From the very beginning, the phenomena discussed under the rubric of symbiosis confronted a view of evolutionary change and ecological relations in terms of ceaseless conflict and competition—a perspective which, critics then and now argue, reflects only an individualist view of human social progress. In this context, debates centered on whether these living arrangements were of mutual benefit to associates, or whether one partner was more or less parasitizing the other. Early in the twentieth century experimentalists who examined the underlying dynamics of such relations argued that they were not fixed but may, at various times in the association, vary from parasitism to mutualism according to environmental conditions. This was as true for the relations of bacteria and heather as it was for the relationship between bacteriophage and bacteria. Some argued that it was difficult if not impossible to demonstrate mutualism; others insisted further that mutualism and parasitism were anthropomorphic categories that have no real existence.

Nonetheless, such categories continued to be the main focus of explanation. To understand why this is so, we have to stand back from the particular phenomenon to be explained and view it in its sociointellectual contexts. We can start where explanation of ecological and organismic evolution meets with explanation of human history. The struggle for existence, taken in its narrow sense as a war of all against all, reinforced a narrow meaning of symbiosis as mutualism. As such, symbiosis as mutualism has been emphasized in opposition to Social Darwinism, during World Wars I and II, during the Cold War, and during the rise of environmentalism in the 1970s. The extent to which symbiosis became equated with mutualism is a measure of the extent to which neo-Darwinism became identified with nature "red in tooth and claw." The ever-continuing attempts to disentangle the concept of symbiosis from an essentialist opposition between mutualism and parasitism reflected the difficulty of removing it from the larger context of Social Darwinism.

One might propose that the dual meaning of symbiosis reflected popular versus technical scientific writing, or that its continued representation in terms of mutualism was due to the effects of extraneous political considerations. However, such a view would be as misleading as the suggestion that the diverse political, philosophical, and religious representations of symbiosis in terms of mutual aid—as finalist, anarchist, communist, internationalist, environmentalist, or feminist—were solely responsible for deterring scientists from studying symbiosis as a means for synthesizing new organisms. After all, even among those ecologists who emphasized cooperation, symbiosis as a means for synthesizing new individuals had always remained in the shadows. This was not simply because underground ecology was largely ignored, but because the cooperative models of ecologists preferred cooperation within species leading to new individuals over cooperation among species.

Nor do I suggest that the significance of symbiosis in evolution will be fully recognized only when it is divorced from any political connotations. Accounts of the origins of symbiotic complexes have always been inextricably interwoven with assumptions about our human nature and debates about our present and future relations with each other and with other species. The acceptance of the symbiotic origin of eukaryotic cells did not preclude this intermingling any more than had discussions of lichens, *Convoluta,* and lupines before it. On the contrary, providing coherent historical reconstructions of events, processes, and contexts was crucial to theorizing about symbiosis.

The units one begins with, the attributes one bestows upon them, and the contexts that bring them together are decisive to the kind of evolutionary story told. If one began with the species, one might see cooperation (at least as an outcome) in contrast to conflict and competition, but if one began with individuals, then cooperation might rely on individual suffering. If microbes were too virulent, then those that were symbiotic had to have a different phylogenetic lineage. If they were cooperative, then it was environmental contexts that made them uncooperative. If they were parasites, then "cooperation" was really a matter of control and domination. If they were neutral, then it was environmental factors which bought them together and maintained the association.

Mutualism versus parasitism was bound up in the question of what counts as an individual—whether, for example, one should classify lichens as parasitic fungi or as a distinct class of organism—and what counted as "progressive evolution"—whether one should classify *Convoluta* as a worm parasitizing an algae or as a plant-animal on its way to becoming a plant. These views themselves were based on the social concept of the individual as a collection of mutually interdependent elementary organisms interacting for the good of the collective whole.

It would be a gross error to reduce the antithesis, mutualism versus parasitism, and what counts as an individual to whimsical reflections of the political attitudes of individual scientists, far removed from actual scientific research and discourse. They are embedded in the aims, concepts, and doctrines that lie deep within the diverse disciplines of the life sciences ever since Darwin. The ever-continuing discussions over the definition of symbiosis—that one could not objectively distinguish mutualism from parasitism, that they were anthropocentric labels, that one had to take into consideration the physiological attributes of partners and their environmental circumstance—were attempts to escape the confines of explanations in terms of contractual relations and conflations of cause and effects, of natural theological finalism on the one hand and neo-Darwinian reductionism on the other. Deflecting the meaning of symbiosis away from fixed conceptions of parasitism and mutualism amounted to removing it from some of the dominant paradigms that have shaped the life sciences in the twentieth century.

The emphasis on master or slave was reinforced by a view of evolution which emphasized the inherent behavioral properties of individuals and their reproductive success. When the outcome of relations among individuals was divorced from understanding contexts, there were only two choices. Individuals were either inherently selfish or inherently altruistic. In the post-Darwinian evolutionary construction, any concept which implied that individuals or populations acted for the common good was confusing ends with means, cause and effect, and was teleological. This opposition, altruism versus selfishness, pervaded discussions of the evolution of biological organization from Kropotkin to sociobiology.

The germ theory of disease further locked discourse about symbiosis into cost-benefit analysis and mutualism. For those who, with Roscoe Pound and August Lumière, emphasized the inherently hostile nature of microbial infections, it was implicit that microbes benefited, but more difficult to accept that plants and animals benefited. They had the opposite problem from de Bary, Frank, Cleveland, Meyer, and many other experimentalists to the present day, who found it difficult to account for the benefit to microbial associates. The meaning of symbiosis was caught in this quandary, from which it still has not escaped.

But whether considered as mutualism or not, there were many other aims and doctrines confronting studies of symbiosis. Inasmuch as they emphasized the creative evolutionary effects of interspecies integration, of microbial "infections," that there was something else beside relations in which the parasite quickly destroys its host, studies of symbiosis remained outside the experimental and conceptual foundations of the modern consensus. First and foremost, the disease-causing aspects of microbes—as "the enemy of Man"—overshadowed their life-giving properties. Second, inasmuch as symbiosis denied a rigorous delimitation of the individual, the study of such associations conflicted with the main thrust of experimental biology with its emphasis on reducing life functions to physics and chemistry. Throughout most of the twentieth century, the view that cells were compounds of other cells was seen as standing in the way of such an understanding—either in an epistemological sense or an operational one. A physicochemical understanding of life meant studying organisms in isolation. Effects due to the presence of microbes were by definition considered to be contamination, error.

This was as true in physiology as it was in genetics during the first half of the century when heredity was defined as the sexual transmission of genes from one generation to the next. Anything else, transmitted sexually or not, was by definition foreign, disease, retrogressive. By definition, species had a common gene pool with reproductive isolating mechanisms that kept them distinct from other species, and immune from their genes, just as organisms themselves had such self-defining mechanisms. Species-specific symbionts were considered to be a rare exceptional occurrence, and of little significance for evolution and heredity. Progressive evolution operated through single genomes. Virtually every organized structure worthy of the neologism "organism" possessed only one genome. By definition, an organized body with mutually connected and dependent parts constituted to share a common life shared one common set of genes, one common inheritance.

The Weismannian view that each organism contained only a single genome was reinforced by Mendelian geneticists' insistence that the nucleus was the sole seat of heredity. The effects of these doctrines on the concern given to symbiosis are well illustrated in the attention given to Portier's theory in France and Wallin's in the United States. In France, where studies of chloroplasts and mitochondria were prevalent, and where studies of microbes in life processes, nutrition, and immunology were emphasized, and where many biologists searched for the effects of the environment on evolutionary change, Portier's *Les Symbiotes* encountered keen interest and intense scrutiny. In the United States, Wallin's *Symbionticism and the Origin of Species* was met with silence.

Yet there were more than these intellectual issues at play in the interest these books captured in their respective countries. We have to consider, for example, the extent to which Portier's application to the chair of physiology at the Sorbonne and to the Muséum d'Histoire Naturelle raised the stakes for his *Les Symbiotes*. We also have to take into account the appreciation in France for large synthetic theories which embraced the collective and broad interests of those botanists and zoologists in Paris who decided upon such appointments.

Herein lies another aspect of the growth of the life sciences which confronted studies of symbiosis in physiology and evolution. The institutionalized boundaries around zoology, botany, bacteriology, virology, genetics, and pathology not only meant that studies pertaining to symbiosis in evolution remained scattered in various journals with disparate interests. It also meant that synthetic studies of symbiosis which crossed such disciplinary lines could not be properly assessed: Minchin, a zoologist, was willing to evaluate nucleus and cytoplasm as symbionts, but he was reluctant to evaluate chloroplasts as symbionts; Caullery was reluctant to discuss bacteria in his book on symbiosis; Portier was judged by his specialist critics to be incompetent to discuss cytology or bacteriology; Wallin's work was said to lie between bacteriology and cytology, where again no one could properly evaluate it. Indeed, it was difficult to control knowledge over symbiosis.

When considering the role of symbiosis in evolution, there was also the issue of its speculative nature. This is not to suggest that symbiosis in evolution, or a microbe-eyed view of evolution, was more speculative than a gene-eyed one. But the latter had already been embedded in the techniques geneticists used, by studying the sexual transmission of differences between individuals within species, and the model organisms they domesticated. Microbial behavior and "underground ecology" had been largely left unconsidered by neo-Darwinian evolutionists, and when it was considered by ecologists there was no consensus over what was to count as an organism any more than there was consensus over what was to count as mutualism.

It was only after World War II, when geneticists extended their studies to include microorganisms, when new evidence for cytoplasmic genetic entities was produced, when they recognized that there were other means besides sex for transmitting genes from one generation to the next, that the idea of broadening the definition of heredity to include infectious entities and broadening the concept of organism to include symbiotic complexes was discussed. As the perceived relationships among intracellular entities graded off between fully integrated genes and infectious entities, and from mutualism to parasitism, so too did the boundaries of the organism.

The acceptance of cytoplasmic entities as part of genetics offered, as Darlington and Lederberg suggested in 1950, a unified theory that would enable one to see the relations among heredity, development, and infection. But the main focus of subsequent genetic work was not on the role of such infections in heredity and evolution, but on their use for studying the nature of the gene, and later tailoring new gene complexes and inserting them into cells. Biologists' ability to insert genes from one species into the genome of another was later interpreted as evidence in favor of the symbiotic nature of chloroplasts and mitochondria.

Throughout the 1960s the main thrust of virology and bacteriology remained tied to studies of disease or, as Dubos remarked, "poor cousins in the mansion of pathology." Attention to the role played by microbes in the early evolution of life on Earth increased during the 1960s and 1970s. Yet only in the study of the origin of the cell itself can one find a sudden shift in the consensus over symbiosis as a source of evolutionary innovation, when studies became amenable to techniques of molecular biology. But here, to better understand what exactly has been accepted, one must be careful in locating that transformation in evolutionary theory and its underlying causes.

Although it is tempting to cast the history of symbiosis in cell evolution in terms of ideas that have been proposed since the late nineteenth century but became accepted only in the late twentieth century, one has to appreciate the differences between early suggestions and the contemporary conceptual consensus. In Altmann's theory, bioblasts were the only living constituents of the cell, a view that was rejected even by symbiosis theorists of his day. Schimper suggested that chloroplasts might be symbionts, but he did not develop the idea. One can also mention Reinke, Haeckel, Schneider, Haberlandt, Watasé, Keeble, and Minchin, among others who discussed the idea favorably in the late nineteenthh and early twentieth centuries.

When Watasé discussed the origin of cells in terms of a symbiosis, his primary focus, and that of many cell and developmental biologists, was on the distinction between nucleus and cytoplasm, which he conceived as being comprised of two phylogenetically distinct elementary organisms. He also mentioned the possibility that centrioles might be symbionts, but he made no mention of mitochondria or bacteria. Similarly, for Famintsyn and Merezhkovskii, the fundamental distinction in all cells was between nucleus and cytoplasm, or mycoplasm and amoeboplasm. They considered chloroplasts also to be symbionts, but Merezhkovskii denied the same for mitochondria.

Portier's emphasis was not on the origin of cells per se, but on the physiological role of symbionts in the life processes of all plants and animals. Mitochondria provided evidence for the universal presence of bacteria in all cells, and for the importance of symbiotic bacteria in nutrition, development, and fertilization. In his scheme, and that of Wallin following him, many cell structures, including chloroplasts, were derived from mitochondria during ontogeny. Wallin's main interest was also not the origin of the cell per se, but to show the importance of symbiotic bacteria for the origin of species, how repeated bacterial symbioses lead to the origin of new genes, and to demonstrate the role of mitochondria/symbionts in developmental processes. His emphasis, like that of Portier before him, was on bacteria as the building blocks of life, and the principal opposition to this view, as he saw it, was the disease-causing attributes of bacteria.

In the genetics of the 1940s and early 1950s, discussions of symbiosis centered around symbiotic viruses as naked genes, as a way of unifying infective heredity in bacteria with evidence for cytoplasmic heredity in other microbes. The exemplar in Lederberg's discussion of symbiosis was lysogeny, not lichens or nitrogen-fixing bacteria. He remained agnostic about the symbiotic origin of chloroplasts and mitochondria. When Dubos discussed symbiosis in evolution, his emphasis, like that of d'Herelle and Wallin before him, was on the importance of symbiosis in the origin of species.

Although these studies and polemics helped pave the way, when symbiosis in evolution did come to the fore, in the 1960s and 1970s, the focus of debate was much more restricted to a specific problem—the role of symbiosis in the origin of eukaryotic cells. There were no great polemics against bacteriologists, or claims for symbiosis in the origin of species. By that time the structural distinction between bacteria and eukaryotes had been accepted. Evidence for the symbiotic origin of cell organelles did not rely solely on gross morphological similarities with bacteria and evidence of symbiotic transitions toward permanent interdependence. DNA had been found to be present in chloroplasts and mitochondria, and genetic research programs on chloroplast and mitochondrial genomes had begun.

Evidence for their symbiotic nature did not rely on culturing them outside the cell. On the contrary, mitochondria and chloroplasts were considered to have lost much of their organismic bacterial nature. Many of their functions are controlled by genes residing in the nucleus. Debate shifted to their symbiotic origin. What remained was to provide plausible accounts, construct new techniques for investigating cell origins, and weigh the evidence against endogenous theories. Only the symbiotic origin of centrioles remains disputed. The symbiotic origin of mitochondria and chloroplasts posed no threat to the major evolutionary consensus regarding the nature of evolutionary change as long as symbiotic associations remained exceptions.

Throughout the 1970s and 1980s, studies of symbiosis in eukaryotic cell evolution were largely removed from studies of its role in other evolutionary transformations. For the most part, investigations of eukaryotic cell evolution were investigated in isolation from those nineteenth-century exemplars which had originally led to speculation on symbiosis in cell evolution. When studies of cooperation came to the fore with the rise of sociobiology, the focus was on cooperation within species with an increased emphasis on individualism in evolutionary explanation. Cooperation was reduced to selfish genes, and kinship theory predicted that mutualism between species would be an exceptional occurrence. While symbiosis researchers now explore the scope of symbiosis as a general feature of evolutionary change, leading neo-Darwinian evolutionists and ecologists continue to assert that symbiosis is not a general characteristic of evolution.

Notes

Introduction

1. For the most recent accounts see Lynn Margulis, *Symbiosis in Cell Evolution. Microbial Communities in the Archean and Proerozoic Eons,* 2nd ed. (New York: W. H. Freeman, 1993). Betsy Dyer and Robert Obar, *The Enigmatic Smile. Tracing the History of Eukaryotic Cells* (New York: Columbia University Press, 1993).

2. Lynn Margulis and René Fester, eds., *Symbiosis as a Source of Evolutionary Innovation* (Cambridge: MIT Press, 1991). Mary Beth Saffo, "Symbiosis in Evolution," in Elizabeth C. Dudley, ed., *The Unity of Evolutionary Biology. Proceedings of the Fourth International Congress of Systematics and Evolutionary Biology* (Portland, Or.: Dioscorides Press, 1991), 674–80.

3. The only exception I have found is a few paragraphs referring to some ideas of Andrei Famintsyn in Alexander Vucinich, *Darwin in Russian Thought* (Berkeley: University of California Press, 1988), 178–79, 182. There are few historical studies which focus on symbiosis; see, however, Douglas Boucher, "The Idea of Mutualism, Past and Future," in D. H. Boucher, ed., *The Biology of Mutualism. Ecology and Evolution* (London: Croom Helm, 1985). For an account of research on symbiogenesis in Russia, see L. N. Khakhina, Lynn Margulis, and Mark McMenamin, eds., *Concepts of Symbiogenesis. History of Symbiosis as an Evolutionary Mechanism* (New Haven: Yale University Press, 1992); see also Donna C. Mehos, "Ivan E. Wallin's Theory of Symbionticism," Appendix, pp. 149–63. Jan Sapp, "Living Together: Symbiosis and Cytoplasmic Inheritance," in Lynn Margulis and René Fester, eds., *Symbiosis as a Source of Evolutionary Innovation* (Cambridge: MIT Press, 1991), 16–25. Jan Sapp, "Symbiosis in Evolution: An Origin Story," *Endocytobiology and Cell Research* 7 (1990): 5–36.

4. Daniel Todes, *Darwin without Malthus. The Struggle for Existence in Russian Evolutionary Thought* (New York: Oxford University Press, 1989).

5. Gregg Mitman, *The State of Nature. Ecology, Community and American Social Thought* (Chicago: University of Chicago Press, 1992).

6. For a brief but useful discussion of the use of "key words" in historical studies, see Evelyn Fox Keller and Elisabeth Lloyd, "Introduction," in E. F. Keller and E. Lloyd, eds., *Keywords in Evolutionary Biology* (Cambridge: Harvard University Press, 1992), 1–6.

7. See Robert Young, *Darwin's Metaphor. Nature's Place in Victorian Culture* (Cambridge: Cambridge University Press, 1985). Camille Limoges, "Milne-Edwards, Darwin, Durkheim and Division of Labour: A Case Study in Reciprocal Conceptual

Exchanges between the Social and Natural Sciences," in I. B. Cohen, ed., *The Relations between the Natural Sciences and the Social Sciences: Historical Interactions* (The Netherlands: Kluwer Academic Publishers, 1994), 317–343. See also Camille Limoges and Claude Ménard, "Organization and Division of Labour: Biological Metaphors at Work in Alfred Marshall's Principles of Economics," in Philp Mirovski, ed., *Natural Images in Economic Thought: Markets Red in Tooth and Claw* (Cambridge: Cambridge University Press, 1994), 336–359.

8. For a discussion of evolutionary narratives in stories of our human origins, see Misia Landau, *Narratives of Human Evolution* (New Haven: Yale University Press, 1991). Bruno Latour and S. C. Strum, "Human Social Origins: Oh Please, Tell Us Another Story," *Journal of Social and Biological Structures* 9 (1986): 169–87. Linnda R. Caporael, "Constructions and Constraints: How to Tell an Evolutionary Tale" in press.

9. David Rindos, *The Origins of Agriculture. An Evolutionary Perspective* (Orlando, Fla.: Academic Press, 1984).

10. Larry Wos and William McCune, "Automated Theorem Proving and Logic Programming: A Natural Symbiosis," *Journal of Logic Programming* 11 (1991): 1–12.

11. Denis Showalter, "Historians and Small Presses: An Emerging Symbiosis?" *Small Press* 8 (1990): 10–14.

Chapter 1

1. Anton de Bary, "Die Erscheinung der Symbiose," in *Vortrag auf der Versammlung der Naturforscher und Ärtze zu Cassel* (Strassburg: Verlag von Karl J. Trubner, 1879): 1–30; 21–22.

2. See, for example, Albert Schneider, *A Textbook of Lichenology* (Lancaster, Pa.: Wiilard N. Clute, 1897), 15–37. Anton de Bary, *Comparative Morphology and Biology of the Fungi Mycetozoa and Bacteria,* trans. Henry E. F. Garnsey, rev. I. B. Balfour (Oxford: Clarendon Press, 1887), 456–59.

3. See Félix Henneguy, *Les Lichens Utiles* (Paris: Masson, 1883).

4. Schneider, *Text-book of Lichenology,* 15–17.

5. Ibid., 19.

6. Simon Schwendener, "Untersuchungen über den Flechtenthallus," *Beiträge zur wissenschaftlichen Botanik* 6 (1968): 195–207; "Die Algentypen der Flechtengonidien," *Programm für die Rectorsfeier der Universität Basel* 4 (1869): 1–42.

7. Quoted in James M. Crombie, "On the Algo-Lichen Hypothesis," *Journal of the Linnaean Society* 21 (1886): 259–282, 261.

8. J. M. Crombie, "On the Lichen-Gonidia Question," *Popular Science Review* 13 (1874): 260–77, 260.

9. Ibid., 276.

10. Ibid., 275.

11. See discussion by Schneider, *Text-book of Lichenology,* 27.

12. For a general account of the institutional and research goals of German botanists of this period, see Eugene Cittadino, *Nature as the Laboratory. Darwinian Plant Ecology in the German Empire, 1880–1900* (Cambridge: Cambridge University Press, 1990). It is interesting to note that, with the exception of one brief mention of lichens, studies of symbiosis are entirely absent from Cittadino's account; the term "symbiosis" is not mentioned at all.

13. For reviews of these studies of lichens, see J. M. Crombie, "On the Algo-Lichen Hypothesis," *Journal of the Linnaean Society* (*Botany*) 21 (1884): 259–83.

Schneider, *Text-book of Lichenology,* 15–37. De Bary, *Comparative Morphology,* 456–459. G. C. Ainsworth, *Introduction to the History of Mycology* (Cambridge: Cambridge University Press, 1976), 97.

14. Crombie, "On the Algo-Lichen Hypothesis."

15. Margaret Lane, *The Tale of Beatrix Po‹‹er* (London: Frederick Warne, 1946), 40–44.

16. E. Bornet, "Recherches sur les gonides des lichens," *Annales des Sciences Naturelles,* 5th ser., 17 (1873): 45–51; "Deuxième note sur les gonidies des lichens," *Annales des Science Naturelles,* 5th ser., 19 (1874): 316–20.

17. See Howard S. Reed, *A Short History of the Plant Sciences* (Waltham, Mass.: Chronica Botanica, 1942), 280. Ainsworth, *History of Mycology,* 98.

18. Ernst Stahl, *Beiträge zur Entwicklungsgescichte der Flechten,* 2 vols. (Leipzig: Arthur Felix, 1877).

19. Gaston Bonnier, "Recherches sur la Synthèse des Lichens," *Annales des Sciences Naturelles Botanique, 7th ser.,* 9 (1989): 1–34.

20. J. Reinke, "Zur Kenntniss des Rhizoms von *Corallorhiza* und *Epipogon,*" *Flora* 31 (1873): 145–209.

21. For a review of the different views on lichen classification during this period, see, for example, Schneider, *Text-book of Lichenology.*

22. A. B. Frank, "Über die biologischen Verhältnisse des Thallus eineger Krustenflechten," *Beitrage zur Biologie der Pflanzen* 2 (1877): 123–200, 195.

23. B. Frank, *Lehrbuch der Botanik* (Leipzig, 1892).

24. See Eugene Cittadino, *Nature as the Laboratory. Darwinian Plant Ecology in the German Empire, 1880–1900* (Cambridge: Cambridge University Press, 1990), 82.

25. De Bary, "Die Erscheinung der Symbiose," 5.

26. De Bary, *Comparative Morphology,* 459.

27. Pierre-Joseph van Beneden, "Un Mot sur la Vie Sociale des Animaux Inférieurs," *Bulletin de l'Académie Royale de Belgique,* 2nd ser., 36 (1873): 779–96.

28. Pierre-Joseph van Beneden, *Animal Parasites and Messmates* (London: Henry S. King, 1876).

29. Ibid., 85.

30. Ibid., 1.

31. Ibid., 53.

32. Ibid., 55.

33. Ibid., 56.

34. Ibid., 68.

35. Ibid., 107.

36. Ibid., 81.

37. De Bary, "Die Erscheinung der Symbiose," 21.

38. Ibid., 11.

39. Ibid., 18.

40. Ibid., 15.

41. Ibid., 21–22.

42. Ibid., 23.

43. Ibid.

44. Ibid., 29.

45. Ibid.

46. Ibid., 29–30.

47. For an account of this research see Paul Buchner, *Endosymbiosis of Animals with Plant Microorganisms* (New York: Interscience, 1965), 3–7.

48. For historical discussion of early challenges to the plant–animal dichtomy, see Lynn Rothschild, "Protozoa, Protista, Protoctista: What's in a Name?" *Journal of the History of Biology* 22 (1989): 277–305. See also F. J. Cole, *The History of Protozoology* (London: University of London Press, 1926).

49. K. Brandt, "Über das Zusammenleben von Algen und Tieren," *Biologisches Centrallblatt* 1 (1881): 524–27.

50. Patrick Geddes, "Further Researches on Animals Containing Chlorophyll," *Nature* 25 (1882): 303–4.

51. Ibid., 304.

52. Ibid., 304–5.

53. Gesa Entz, "Über die Natur der Chlorophyllkörperchen nieder Tiere," *Biologisches Centralblatt* 1 (1881): 646–50. Originally published as "Az alsóbb rangú állatoknál elöforduló levélzöldtestecskék természetéröl," *Kolozsvári Orvos-természettudományi Ertesítö,* 25 February 1876.

54. See letters to *Nature* 25 (1882): 361–62.

55. For an overview of the research on "animal chlorophyll" and symbiosis, see E. L. Bouvier, "La Chlorophylle Animale et les Phénomènes de Symbiose entre les Algues Vertes Unicellulaires et les Animaux," *Bulletin de la Société Philomathique de Paris* 5 (1893): 72–149. See also John F. Fulton, "Animal Chlorophyll: Its Relation to Haemoglobin and to Other Animal Pigments," *Quarterly Journal of Microscopical Science* 66 (1922): 340–96.

56. A. Famintzin, "Nochmals die Zoochlorellen," *Biologisches Centralblatt* 12 (1892): 51–54.

57. For an account of the research on nitrogen fixation, see P. Wilson and E. B. Fred, "The Growth Curve of a Scientific Literature: Nitrogen Fixation by Plants," *Scientific Monthly* 41 (1935): 240–50.

58. J. R. Green, *A History of Botany 1860–1900* (Oxford: Clarendon Press, 1909). Howard S. Reed, *A Short History of the Plant Sciences* (Waltham, Mass.: Chronica Botanica, 1942).

59. A. B. Frank, "Über die auf Wurzelsymbiose beruhende Ernährung gewisser Bäume durch unterirdische Pilze," *Berichte der Deutschen Botanischen Gesellschaft* 3 (1885): 128–45.

60. G. C. Ainsworth, *Introduction to the History of Mycology* (Cambridge: Cambridge University Press, 1976), 100–113.

61. Ibid., 104–5.

62. Noël Bernard, "Études sur la Tubérisation," *Revue Génerale Botanique* 14 (1902): 5–25; "L'Evolution dans la symbiose. Les orchidée et leur champignons commensaux," *Annales des Sciences Naturelles de Paris, 9th ser.,* 9 (1909): 1–196; "Remarques sur l'immunité chez les plantes," *Bulletin de l'Institut Pasteur* 12 (1909): 369–86.

63. Bernard, "Remarques sur l'immunité," 377.

64. See Maurice Caullery, *Parasitism and Symbiosis* (London: Sidgwick and Jackson, 1952), 266–75.

65. De Bary, *Comparative Morphology,* 369.

Chapter 2

1. Roscoe Pound, "Symbiosis and Mutualism," *American Naturalist* 27 (1893): 509–20, 519.

2. See, for example, John Greene, "Biology and Social Theory in the Nineteenth Century," in Marshall Clagett, ed., *Critical Problems in the History of Science* (Madison: University of Wisconsin Press, 1962), 416–46. Richard Hofstadter, *Social Darwinism in American Thought* rev. ed. (Boston: Beacon Press, 1955). Greta Jones, *Social Darwinism in English Thought* (London: Harvester, 1980). Robert Young, *Darwin's Metaphor. Nature's Place in Victorian Culture* (Cambridge: Cambridge University Press, 1985); "Darwinism Is Social," in David Kohn, ed., *The Darwinian Heritage* (Princeton: Princeton University Press, 1985), 609–38. For a critique of Young's approach, see Ingemar Bohlin, "Robert M. Young and Darwin Historiography," *Social Studies of Science* 21 (1991): 597–648. Jim Moore, "Socializing Darwinism: Historiography and the Fortunes of a Phrase," in Les Levidow, ed., *Science as Politics* (London: Free Association Books, 1986), 38–75.

3. Charles Darwin, *On the Origin of Species by Means of Natural Selection* (London: John Murray, 1859). Reprinted with an introduction by Ernst Mayr (Cambridge: Harvard University Press, 1966), 62.

4. Ibid., 92–93.

5. Ibid., 63.

6. See Robert Young, "Malthus and the Evolutionists: The Common Context of Biological and Social Thought," in *Darwin's Metaphor,* 23–55.

7. Ibid., 25.

8. Ibid., 63.

9. Ibid., 75.

10. Darwin, *The Origin,* 490.

11. For a discussion of Proudhon's mutualism and its relation to the biological concept, see D. H. Boucher, "The Idea of Mutualism, Past and Future," in D. H. Boucher, ed., *The Biology of Mutualism. Ecology and Evolution* (London: Croom Helm, 1983), 1–28, 12–13.

12. See E. Schulkind, ed., *The Paris Commune of 1871: The View from the Left* (London: Jonathan Cape, 1972), 164; quoted in Boucher, "The Idea of Mutualism," 12.

13. Boucher, "The Idea of Mutualism," 12.

14. P. J. van Beneden, *Animal Parasites and Messmates* (London: Henry S. King, 1876), xvii.

15. Boucher, "The Idea of Mutualism," 14.

16. Ibid.

17. A. Kemna, *P. J. van Beneden. La vie et l'oeuvre d'un Zoologiste* (Anvers: J.-E. Buschmann, 1897), 20.

18. Van Beneden, *Animal Parasites,* xxvii.

19. See Frank Egerton, "Changing Concepts of the Balance of Nature," *Quarterly Review of Biology* 48 (1973): 322–50, 326.

20. Ibid., 329.

21. Ibid., 186.

22. Ibid.

23. Quoted in Donald Worster, *Nature's Economy. A History of Ecological Ideas* (Cambridge: Cambridge University Press, 1977), 37.

24. Ibid. See also Carolyn Merchant, *The Death of Nature. Women, Ecology and the Scientific Revolution* (New York: Harper and Row, 1983).

25. See Worster, *Nature's Economy,* 45.

26. Van Beneden, *Animal Parasites,* xviv.

27. Ibid., 29.

28. Kemna, *P. J. van Beneden,* 132.
29. A. Espinas, *Des Sociétés Animales* (Paris: Bailliere, 1878).
30. See Roger Baldwin, ed., *Kropotkin's Revolutionary Pamphlets* (New York: Dover Publications, 1970). Martin A. Miller, *Kropotkin* (Chicago: University of Chicago Press, 1976).
31. Baldwin, *Kropotkin's Revolutionary Pamphlets,* 20–25.
32. Quoted in Peter Kropotkin, *Mutual Aid. A Factor of Evolution* (London: William Heinemann, popular edition, 1915), 13.
33. T. H. Huxley, *Evolution and Ethics and Other Essays* (New York: Columbia University Press, 1911).
34. Kropotkin, *Mutual Aid,* 2.
35. Ibid., 3.
36. Daniel P. Todes, *Darwin without Malthus. The Struggle for Existence in Russian Evolutionary Thought* (New York: Oxford University Press, 1989), 168.
37. Kropotkin, *Mutual Aid,* 13.
38. Ibid., 62.
39. Ibid., 20.
40. Ibid., 14.
41. Ibid., 17.
42. Quoted in Todes, *Darwin without Malthus,* 115.
43. Ibid., 117.
44. Patrick Geddes and J. Arthur Thomson, *The Evolution of Sex* (London: Walter Scott, 1889), 290.
45. Ibid., 312.
46. Ibid., 279.
47. Ibid., 310.
48. Ibid., 311.
49. Ibid.
50. Patrick Geddes and J. Arthur Thomson, *Evolution* (London: Williams and Norgate, 1911), 175–76, 246.
51. Ibid., 175.
52. See Hofstadter, *Social Darwinism,* 96–97.
53. Henry Drummond, *The Ascent of Man* (New York: James Pott, 1894), 36.
54. Ibid., 242.
55. Ibid., 232.
56. Ibid., 233.
57. Ibid., 240.
58. Ibid., 234.
59. Ibid., 238–41, 241.
60. See, for example, Hofstadter, *Social Darwinism.*
61. See, for example, Robert Bannister, *Social Darwinism: Science and Myth in Anglo-American Social Thought* (Philadelphia: Temple University Press, 1979); Howard L. Kaye, *The Social Meaning of Modern Biology. From Social Darwinism to Sociobiology* (New Haven: Yale University Press, 1986), 26–31; ibid.
62. Herbert Spencer, *The Principles of Biology,* vol 2, rev. and enl. ed. (New York: D. Appleton and Co., 1899), 533.
63. Ibid.
64. See Hofstadter, *Social Darwinism,* 80.
65. Herbert Spencer, *The Principles of Ethics* (New York: D. Appleton and Co., 1892), 189.

66. Ibid., 201.

67. Spencer, *The Principles of Biology,* 2:vi.

68. Ibid., 2:408.

69. It is interesting to note that Worster, who mentions this chapter of Spencer's *Principles of Biology,* does not mention symbiosis. Instead, he asserts that Spencer had simply plucked examples of mutual aid from "Darwin's web of life." See Worster, *Nature's Economy,* 213.

70. Camille Limoges, "Milne-Edwards, Darwin, Durkheim and Division of Labor: A Case Study in Reciprocal Conceptual Exchanges between the Social and Natural Sciences," in I. B. Cohen, ed., *The Relations between the Natural Sciences and the Social Sciences* (The Netherlands: Kluwer Academic Publishers, 1994, pp. 317–343). See also Camille Limoges and Claude Ménard, "Organization and Division of Labour: Biological Metaphors at Work in Alfred Marshall's Principles of Economics," in Mirovski ed., *Natural Images in Economic Thought,* 336–359.

71. Darwin, *The Origin,* 115.

72. Spencer, *Principles of Biology,* 396.

73. Ibid., 398.

74. Ibid.

75. Ibid.

76. Ibid., 399.

77. Ibid.

78. Ibid., 400.

79. Ibid., 401.

80. Ibid.

81. Ibid., 405.

82. Ibid., 405–6.

83. Ibid., 408.

84. See, for example, Worster, *Nature's Economy.*

85. Frederick Clements, *Plant Succession. An Analysis of the Development of Vegetation* (Washington: Carnegie Institution of Washington, 1916), 3.

86. Eugenius Warming, *Oecology of Plants. An Introduction to the Study of Plant Communities* (Oxford: Clarendon Press, 1909), 95.

87. Ibid., 349.

88. Worster, *Nature's Economy,* 198.

89. Warming, *Oecology of Plants,* 84.

90. Ibid., 85.

91. Ibid., 83.

92. Ibid., 349.

93. Ibid., 95.

94. See Ronald C. Tobey, *Saving the Prairies. The Life Cycle of the Founding School of American Plant Ecology, 1885–1955* (Berkeley: University of California Press, 1981), 20, 77. Worster, *Nature's Economy,* 209.

95. S. Glueck, ed., *Roscoe Pound and Criminal Justice* (Dobbs Ferry, N.Y.: Oceana Publications, 1965), 65. Quoted in Boucher, "The Idea of Mutualism," 16.

96. Pound, "Symbiosis and Mutualism," 509.

97. Ibid.

98. Ibid., 512.

99. Ibid., 514.

100. Ibid., 513.

101. Ibid., 515.

102. Ibid., 516.
103. Ibid., 519.
104. Darwin, *The Origin,* 10, 11, 13, etc.
105. Pound, "Symbiosis and Mutualism," 519.
106. Albert Schneider, *A Text-book of Lichenology* (Lancaster, Pa.: Willard N. Clute, 1897).
107. Albert Schneider, "The Phenomena of Symbiosis," *Minnesota Botanical Studies* 1 (1897): 923–48.
108. Ibid., 925.
109. Ibid., 926–27.
110. Ibid., 927.
111. Ibid., 930.
112. Ibid., 929.
113. Ibid., 937.
114. Ibid., 936.
115. Ibid., 926.
116. Ibid., 936.
117. Ibid., 940.
118. Ibid.
119. Ibid., 942.
120. Ibid., 940.
121. Albert Schneider, "Mutualistic Symbiosis of Algae and Bacteria with *Cycas revoluta,*" *Botanical Gazette* 19 (1894): 25–32.
122. Schneider, "The Phenomena of Symbiosis," 943.
123. Ibid., 944.

Chapter 3

1. E. B. Wilson, *The Cell in Development and Inheritance* (New York: Macmillan, 1896), 211.
2. See Yves Delage, *La Structure du Protoplasm et Les Théories sur l'Heredité et les Grands Problèmes de la Biologie Générale* (Paris: C. Reinwald and Co., 1895). Wilson, *The Cell,* 1896.
3. L. S. Jacyna, "Romatic Thought and the Origins of Cell Theory," in A. Cunningham and N. Jardine, eds., *Romanticism and the Sciences* (Cambridge: Cambridge University Press, 1990), 161–68.
4. Paul Weindling, "Theories of the Cell State in Imperial Germany," in Charles Webster, ed., *Biology, Medicine and Society 1840–1940* (Cambridge: Cambridge University Press, 1981), 99–154.
5. Herbert Spencer, "Professor Weismann's Theories," *Contemporary Review* 63 (1893): 743–60.
6. Weindling, "Theories of the Cell State," 100.
7. Ibid., 114.
8. See Jane Maienschein, "Whitman at Chicago: Establishing a Chicago Style of Biology?" in Keith R. Benson and Jane Maienschein, eds., *The American Development of Biology* (Philadelphia: University of Pennsylvania Press, 1988), 151–84; *100 Years Exploring Life, 1888–1898. The Marine Biological Laboratory at Woods Hole* (Boston: Jones and Bartlett, 1989).
9. C. O. Whitman, "Socialization and Organization, Companion Principles of All

Progress. The Most Important Need of American Biology," *Biological Lectures Delivered at the Marine Biological Laboratory of Woods Hole in the Summer Session of 1890* (Boston: Ginn and Co., 1891), 1–26.

10. Ibid., 19.

11. Ibid., 21.

12. Ibid., 6.

13. Ibid., 25.

14. Wilson, *The Cell,* 41.

15. Ibid.

16. E. B. Wilson, *The Cell in Development and Heredity* (New York: Macmillan, 1925), 102.

17. For a historical account of the development of this view, see William Coleman, "Cell, Nucleus, and Inheritance: An Historical Study," *Proceedings of the American Philosophical Society* 109 (1965): 124–58.

18. Richard Altmann, *Die Elementarorganismen* (Leipzig: Veit, 1890).

19. See Wilson, *The Cell* (1896), 21.

20. Delage, *La Structure du Protoplasm,* 502–5.

21. See Alexandre Guilliermond, *The Cytoplasm of the Plant Cell* (Waltham, Mass.: Chronica Botanica, 1941), 56–84.

22. Wilson, *The Cell* (1896), 53–56.

23. For a recent discussion of the important role of centrosomes in cell division, motility, and shape, see David M. Glover, Cayetano Gonzalez, and Jordan W. Raff, "The Centrosome," *Scientific American* 266 (1993): 62–68.

24. A. F. W. Schimper, "Untersuchungen über die Chlorophyllkörner und die ihnen homologen Gebilde," *Jahrbücher für wissenschaftliche Botanik* 16 (1885): 1–247, 202.

25. Cited in A. F. W. Schimper, "Über die Entwicklung der Chlorophyllkörner und Farbkörper," *Botanische Zeitung* 41 (1883): 105–14, 112–13.

26. Ibid., 105–14, 121–31, 137–46, 153–60, quote from 112–13.

27. Donald Worster, *Nature's Economy. A History of Ecological Ideas* (Cambridge: Cambridge University Press, 1977), 199. Eugene Cittadino, *Nature as the Laboratory. Darwinian Plant Ecology in the German Empire, 1880–1900* (Cambridge: Cambridge University Press, 1990).

28. Frederick Keeble, *Plant-Animals. A Study in Symbiosis* (Cambridge: Cambridge University Press, 1910).

29. Gottlieb Haberlandt, "Über den Bau und die Bedeutung der Chlorophyllzellen von *Convoluta roscoffensis,*" in Ludwig von Graff, *Die Organization der Turbellaria Acoela* (Leipzig: Verlag von Wilhelm Engelmann, 1891), 75–90.

30. F. R. Lillie and Sigeo Yamanouchi, "Shôsaburô Watasé", *Science* 71 (1930): 577–78.

31. Ibid.

32. S. Watasé, "Homology of the Centrosome," *Journal of Morphology* 8 (1893): 433–43.

33. Lillie and Yamanouchi, "Shosaburo Watasé," 578.

34. S. Watasé, "On the Nature of Cell Organization," *Woods Hole Biological Lectures,* 1893, 83–103, 89.

35. Ibid., 92.

36. Ibid., 90.

37. Ibid., 86.

38. Ibid., 93.

39. The differentiation hypothesis was maintained, for example, by Max Verworn, *Die Physiologische Bedeutung des Zell-kerns* (Bonn, 1891), 115. Julius Wiesner, *Die Elementarstructure und das Wachsthum der lebenden Substanz* (Vienna, 1892), 266.

40. Watasé, "On the Nature of Cell Organization," 101.

41. Verworn, *Die Physiologische Bedeutung,* 42.

42. Watasé, "The Nature of Cell Organization," 96.

43. Ibid., 103.

44. Ibid.

45. See, for example, the views of Tswett, 1896, in A. Famintsyn, "Die Symbiose als Mittel der Synthese von Organismen," *Biologisches Centralblatt* 27 (1907): 352–63, 360.

46. Hugo de Vries, *Intracellular Pangenesis,* trans. J. B. Farmer and A. D. Darbishire (Chicago: Open Court, 1910).

47. August Weismann, *The Germ Plasm: A Theory of Heredity,* trans. W. N. Parker and H. Rofeldt (London: Walter Scott, 1893), 22.

48. Ibid., 23.

49. Ibid., 27.

50. Ibid., 29.

51. Ibid., 26.

52. Ibid., 49.

53. See, for example, F. B. Churchill, "August Weismann and a Break from Tradition," *Journal of the History of Biology* 1 (1968): 91–112.

54. Wilson, *The Cell* (1896), 327.

55. Ibid., 37.

56. Ibid., 211–12.

Chapter 4

1. C. Mérejkovsky, "La Plante considérée comme un complexe symbiotique," *Bulletin de la Société Naturelles* 6 (1920): 17–98, 17.

2. Lynn Margulis, "Symbiogenesis and Symbionticism," in Lynn Margulis and René Fester, eds., *Symbiosis as a Source of Evolutionary Innovation* (Cambridge: MIT Press, 1991), 1–14, 1.

3. E. B. Wilson, *The Cell in Development and Heredity* (New York: Macmillan, 1925), 738.

4. E. A. Minchin, "The Evolution of the Cell," *Report of the Eighty-Fifth Meeting of the British Association for the Advancement of Science,* September 7–11 (1915): 437–64. This paper was published the next year in *American Naturalist* 50 (1916): 3–38, 106–18, 271–83.

5. Lilya Nikolaevna Khakhina, *Concepts of Symbiogenesis,* ed. Lynn Margulis and Mark McMenamin, trans. Stephanie Merkel and Robert Coalson, foreword Alexander Vucinich (New Haven: Yale University Press, 1992).

6. See, for example, Mark B. Adams, "Through the Looking Glass: The Evolution of Soviet Darwinism," in Hilary Koprowski and Leonard Warren, eds., *New Perspectives on Evolution* (New York: Wiley-Liss, 1991), 37–63.

7. Alexander Vicunich, *Darwin in Russian Thought* (Berkeley: University of California Press, 1988).

8. See, for example, A. Famintsyn, "Die Symbiose einer *Chaetoceras*-Art mit

einer Protozoe (*Tintinnus inquilinus*)," *Mémoire de l'Académie de St. Petersburg,* 7th ser., 36 (1889): 1–36; "Nochmals die Zoochlorellen," *Biologisches Centralblatt* 12 (1892: 51–54.

9. See, for example, Maurice Caullery, *Parasitism and Symbiosis,* trans. Averil M. Lysaght (London: Sidgwick and Jackson, 1952), 276.

10. See Vucinich, *Darwin in Russian Thought,* 178–83.

11. Ibid.

12. A. S. Famintzin, "Die Symbiose also Mittel der Synthesis von Organismen," *Biologisches Centralblatt* 27 (1907): 253–64.

13. Ibid., 255.

14. Ibid.

15. Ibid.

16. Andrei Famintsyn, "Beitrag zur Symbiose von Algen und Thieren," *Mémoire de l'Académie des Sciences, St. Petersbourg,* 7th ser., 26 pt. 1 (1889) 36 pp.; pt. 2 38 (1891) 16 pp.

17. Famintzin, "Die Symbiose," 261.

18. Ibid., 362.

19. Ibid., 362–63.

20. Andrei Famincyn, "Note sur les Bryopsis de la côte de Monaco," *Bulletin de l'Institut Océanographique,* no. 200, 10 March, 1911, 1–3.

21. Andrei Famintsyn, "The Role of Symbiosis in the Evolution of Organisms," *Bulletin de l'Académie Impériale des Sciences de St. Petersbourg,* 6th ser., 11 (1912): 51–68, 60.

22. Ibid., 63.

23. Ibid., 66.

24. See Jane Maienschein, "Experimental Biology in Transition: Harrison's Embryology, 1895–1910," *Studies in History of Biology* 6 (1983): 107–27.

25. Ibid.

26. Quoted in Mérejkovsky "La Plante considérée comme un complexe symbiotique," 65.

27. Ibid., 18.

28. C. Mereschkovsky, "Über Natur und Ursprung der Chromatophoren im Pflanzenreiche," *Biologisches Centralblatt* 25 (1905): 595–604.

29. E. Ray Lankester, "Animal Chlorophyll," *Nature* 44 (1891): 344–345.

30. Ibid., 19.

31. For a detailed review of Haeckel's concept of Monera, see Minchin, "The Evolution of the Cell."

32. Philip F. Rehbock, "Huxley, Haeckel, and the Oceanographers: The Case of *Bathybius haeckelii,*" *Isis* 66 (1976): 504–53.

33. See Gerald L. Geison, "The Protoplasmic Theory of Life and the Vital-Mechanist Debate," *Isis* 60 (1969): 273–92.

34. C. Mereschkowsky, "Theorie der zwei Plasmaarten als Grundlage der Symbiogenesis, einer neuen Lehre von der Entstehung der Organismen," *Biologisches Centralblatt* 30 (1910): 277–303, 321–47, 353–67.

35. Theodor Boveri, "Ergennisse über die Konstitution der chromatischen Kernsubstance," quoted in E. B. Wilson, *The Cell in Development and Heredity* (New York: Macmillan, 1925), 1108.

36. Mérejkovsky, "La Plante considérée comme un complexe symbiotique," 19–20.

37. Ibid., 20.

38. Ibid., 20–21.

39. Y. Delage and E. Hérouard, *La Cellule et les Protozoaires. Traité Zoologie Concrète,* vol. 1 (Paris: Schleicher Frères, 1896).
40. Mérejkovsky, "La Plante considérée comme un complexe symbiotique," 24.
41. Ibid., 26.
42. Ibid., 27.
43. Ibid., 28.
44. Ibid., 30–31.
45. Merezhkovskii cites Boveri's paper. Ibid., 75.
46. In fact, Merezhjkovskii mentioned Famintsyn only once, in a footnote in his paper of 1905, when he pointed out that in 1889 Famintsin had studied the independent nature of chromatophores and believed it was possible that one might be able to cultivate them outside of the cell. "Über Natur und Ursprung," 559.
47. Mérejkovsky, "La Plante considérée comme un complexe symbiotique," 39.
48. Ethel Rose Spratt, "Some Observations on the Life-history of *Anabeana cycadeae,*" *Annals of Botany* 25 (1911): 369–80, 277–378, also quoted in Mérejkovsky, "La Plante considérée comme un complexe symbiotiques," 81.
49. Mérejkovsky, "La Plante considérée comme un complexe synbiotique," 82.
50. Ibid., 56–59.
51. Ibid., 60–62.
52. Ibid., 62.
53. Ibid.
54. Ibid., 98.

Chapter 5

1. E. A. Minchin, "The Evolution of the Cell," *Report of the Eighty-Fifth Meeting of the British Association for the Advancement of Science,* September 7–11 (1915): 437–64, 456.
2. Peter Kropotkin, *Mutual Aid. A Factor of Evolution* (London: William Heinemann, popular edition, 1915), v.
3. Ibid., vi.
4. Hermann Reinheimer, *Symbiogenesis. The Universal Law of Progressive Evolution* (Westminister: Knapp Drewett and Sons, 1915).
5. Hermann Reinheimer, *Evolution by Co-operation. A Study in Bio-economics* (Westminister: Knapp Drewett and Sons, 1913).
6. Reinheimer, *Symbiogenesis.*
7. Henry Drummond, *The Ascent of Man* (New York: James Pott, 1894).
8. Peter Bowler, *The Eclipse of Darwinism. Anti-Darwinian Evolution Theories in the Decades around 1900* (Baltimore: Johns Hopkins University Press, 1983).
9. G. H. Theodor Eimer, *Organic Evolution as the Result of the Inheritance of Acquired Characters According to the Laws of Organic Growth,* trans. J. T. Cunningham (London: Macmillan, 1890).
10. Ibid., 427.
11. Ibid., 434.
12. Ibid., 434.
13. Henri Bergson, *Creative Evolution,* trans. Arthur Mitchell (New York: H. Holt, 1911).
14. Reinheimer, *Symbiogenesis,* 382.
15. Ibid., 381.

16. Ibid., 385.
17. Reinheimer, *Evolution by Co-operation,* 1.
18. Ibid., 24.
19. Ibid., 17.
20. Ibid., 18.
21. Ibid., 22.
22. Reinheimer, *Symbiogenesis,* xiv.
23. Reinheimer, *Evolution by Co-operation,* 19.
24. Ibid., 153.
25. Ibid., 154.
26. Ibid., 161.
27. Ibid., 162.
28. Reinheimer, *Symbiogenesis,* 3.
29. Ibid., xiii.
30. Ibid., xvii.
31. Ibid., xix.
32. Ibid., 14.
33. Frederick Keeble, *Plant-Animals. A Study in Symbiosis* (Cambridge: Cambridge University Press, 1910).
34. Ibid., 82.
35. Ibid., 42.
36. Ibid., 147.
37. Ibid., 129.
38. Ibid., 114.
39. Ibid.
40. Ibid., 106.
41. Ibid., 114.
42. Ibid., 131.
43. Ibid., 153.
44. Ibid., 157.
45. Reinheimer, *Symbiogenesis,* 38.
46. Ibid., 69.
47. Ibid., 71.
48. Ibid.
49. Ibid., 339.
50. Ibid., 438.
51. Ibid.
52. E. A. Minchin, *An Introduction to the Study of the Protozoa* (London: Edward Arnold, 1912).
53. See John O. Corliss, "A Salute to Fifty-Four Great Microscopists of the Past: A Pictorial Footnote to the History of Protozoology," *Transactions of the American Microscopic Society* 97 (1978): 419–58.
54. See Natasha Jacobs, "From Unity to Unity: Protozoology, Cell Theory, and the New Concept of Life," *Journal of the History of Biology* 22 (1989): 215–42.
55. See Marsha L. Richmond, "Protozoa as Precursors of Metazoa: German Cell Theory and Its Critics at the Turn of the Century," *Journal of the History of Biology* 22 (1989): 243–76.
56. Richard Hertwig, "Die Protozoen und die Zelltheorie," *Archiv für Protistenkunde* 1 (1902): 1–40.
57. Minchin, "The Evolution of the Cell," 439.

58. T. H. Huxley, "The Cell Theory," *British and Foreign Medico-Chirurgical Review* 12 (1853): 221–243, 243.

59. Julius Sachs, "Lectures on Physiology," quoted in C. O. Whitman, "The Inadequacy of the Cell Theory of Development," *Journal of Morphology* 8 (1893): 639–58, quote on p. 641.

60. Whitman, "Cell Theory," 653.

61. Adam Sedgwick, "On the Inadequacy of the Cellular Theory of Development, and on the Early Development of Nerves, Particularly of the Third Nerve and of the Sympathetic in Elasmobranchii," *Quarterly Journal of Microscopic Science* 37 (1895): 87–101.

62. See for example, Gary Calkins, *The Biology of Protozoa* (Philadelphia: Lea and Febiger, 1933), 19. Richmond, "Protozoa as Precursors." John Corless, "The Protozoon and the Cell: A Brief Twentieth Century Overview," *Journal of the History of Biology* 22 (1989): 307–23.

63. C. Clifford Dobell, "The Principles of Protistology," *Archiv für Protistenkunde* 23 (1911): 269–310.

64. Ibid., 270.

65. Ibid., 284–85.

66. Ibid., 302–3.

67. Minchin, "The Evolution of the Cell," 437.

68. Ibid., 463.

69. Ibid.

70. Ibid., 439.

71. Ibid., 445.

72. Ibid., 463.

73. Ibid., 448.

74. Ibid., 462.

75. Ibid., 449.

76. Ibid., 447.

77. Ibid., 453.

78. Ibid.

79. Ibid., 454.

80. Ibid., 456.

81. Ibid., 457.

82. For a historical account of discussions of chromidia, see Richmond, "Protozoa as Precursors."

83. Minchin, *An Introduction to the Study of Protozoa,* 66.

84. Minchin, "The Evolution of the Cell," 460.

85. Ibid., 461.

86. In the third edition of *The Cell in Development and Heredity* (New York: Macmillan, 1925), 733–39, E. B. Wilson added a little section on "Historical Problems of the Cell" and acknowledged his indebtedness to Minchin.

87. Carl Benda, "Neue Mitteilungen über die Histogenese der Säugetierspermatozoen," *Verhandlungen der Physiologische Gesellschaft* (1897): 406–14. For one of the most extensive early reviews of the literature pertaining to mitochondria research, see E. V. Cowdry, "The Mitochondrial Constituents of Protoplasm," in *Contributions to Embryology,* Carnegie Institution Publications No. 25 (1918): 41–160. Altmann is still today considered to be the discoverer of mitochondria; for a recent review see Lars Enster and Gottfried Schatz, "Mitochondria: A Historical Review," *Journal of Cell Biology* 91 (1981): 227s–255s.

88. See, for example, Cowdry, "The Mitochondial Constituents," 18; J. Duesberg, "Chondriosomes in the Cells of Fish Embryos," *American Journal of Anatomy* 21 (1917): 465–96; Raphaël Dubois, "Symbiotes, Vacuolides, Mitochondries et Leucites," *Comptes rendus de la Société de Biologie* 92 (1919): 473–77.

89. Cowdry, "The Mitchondrial Constituents," 42.

90. For a fuller discussion of "technique-ladenness of observations" as it pertains to the cytoplasmic–nuclear debate, and in genetics generally, see Jan Sapp, "Inside the Cell: Genetic Methodology and the Case of the Cytoplasm," in J. A. Schuster and R. R. Yeo, eds., *The Politics and Rhetoric of Scientific Method* (Dordrecht: Reidel, 1986), 167–202; *Cytoplasmic Inheritance and the Struggle for Authority in Genetics* (New York: Oxford University Press, 1987); *Where the Truth Lies. Franz Moewus and the Origins of Molecular Biology* (New York: Cambridge University Press, 1991).

91. E. V. Cowdry, "The General Functional Significance of Mitochondria," *American Journal of Anatomy* 19 (1916): 423–46, 442.

92. Friedrich Meves, "Die Chondriosomen als Träger erblicher Anlagen. Cytologische Studien am Hühner-embryo," *Archiv für Mikroscopische Anatomie* 72 (1908): 816–67, 845. Meves developed this view further in his "Die Plastosomentheorie der Vererbung. Eine Antwort auf verschieden Einwände," *Archiv für Mikroscopische Anatomie* 92 (1918): 42–136.

93. Claude Regaud, "Attribution aux 'formation mitochondriales' de la fonction générale d'extraction et de fixation électives, exercée par les cellules vivantes sur les substances dissoutes dans le milieu ambiant," *Comptes rendus de la Société Bioloqique* 66 (1909): 919–12.

94. For a fuller discussion of these issues, See Sapp, *Beyond the Gene.*

95. Cowdry, "The Mitochondrial Constituents," 96.

96. C. Mérejkovsky, "La Plante considérée comme un complexe symbiotique," *Bulletin de la Société des Sciences Naturelles de L'ouest de la France* 6 (1920): 17–98, 84.

Chapter 6

1. Paul Portier, *Les Symbiotes* (Paris: Masson, 1918), vii.

2. See John Farley, *The Spontaneous Generation Controversy from Decartes to Oparin* (Baltimore: Johns Hopkins University Press, 1977). Bruno Latour, *The Pasteurization of France,* trans. Alan Sheridan and John Law (Cambridge, Mass.: Harvard University Press, 1988).

3. Interestingly enough, there is no mention of Portier's *Les Symbiotes* in the account of his work in *The Dictionary of Scientific Biography.*

4. Camille Limoges, "A Second Glance at Evolutionary Biology in France," in Ernst Mayr and William B. Provine, eds., *The Evolutionary Synthesis. Perspectives on the Unification of Biology.* (Cambridge, Mass.: Harvard University Press, 1980), 322–28.

5. Ibid., 325.

6. See Emile Guyénot, "Hérédité et variation," *Revue générale des Sciences* 29 (1918): 302–6; "Lamarckism ou Mutationnisme," *Revue générale des Sciences* 32 (1921): 598–606.

7. Emile Guyénot, *L'Hérédité* (Paris: Doin, 1924).

8. Noël Bernard, "Etudes sur la Tubérisation," *Revue Générale Botanique* 14 (1902): 5–25, 5.

9. Ibid., 7.

10. Ibid., 9.

11. Ibid.

12. Noël Bernard, "Remarques sur l'immunité chez les plantes," *Bulletin de l'Institut Pasteur* 12 (1909): 369–86, 370.

13. See, for example, W. Bulloch, *The History of Bacteriology* (New York: Oxford University Press, 1938). H. Lechevalier and M. Solotorovsky, *Three Centuries of Microbiology* (New York: McGraw-Hill, 1965).

14. Alfred I. Tauber and Leon Chernyak, *Metchnikoff and the Origins of Immunology* (New York: Oxford University Press, 1991), 103.

15. Ibid. See also Daniel Todes, *Darwin without Malthus. The Struggle for Existence in Russian Evolutionary Thought* (New York: Oxford University Press, 1989).

16. Bernard, "Remarques sur l'immunité chez les plantes," 377.

17. Ibid., 372.

18. Ibid., 371.

19. Ibid.

20. Ibid.

21. Ibid., 372.

22. On the discovery of anaphylaxis, see M. Fountaine, "La Découverte de l'Anaphylaxie," *Bulletin de l'Institut Océanographique,* no. 997 (1950): 1–9. L. Binet, "L'Evolution de nos connaissances sur l'Anaphylaxie," *Bulletin de l'Institut Océanographique,* no. 997 (1950): 11–18. Charles D. May, "The Ancestry of Allergy: Being an Account of the Original Experimental Induction of Hypersensitivity Recognizing the Contribution of Paul Portier," *Journal of Allergy and Clinical Immunology* 75 (1985): 485–95.

23. Eric Mills and Jacqueline Carpine-Lancre, "The Musée Océanographique," in press.

24. Ibid.

25. For biographical information on Portier, see Paul Courrier, "Notice sur la vie et l'oeuvre de Paul Portier," Archives de Musée Océanographique, Paris, Palais de l'Institut, 1966. M. Bernard Massé, "Paul Portier, 1866–1962, sa vie—son oeuvre," *Thèse pour le Doctorat en Médecine,* Faculté de Médecine de Paris, 1969.

26. May, "The Ancestry of Allergy," 490.

27. Ibid., 493.

28. Paul Portier, "Rapport de M. P. Portier, Laboratoire de Physiologie des Etres Marins," in *Rapport de Different Services de l'Etablissement de Paris, et de Musée de Monaco,* Archives de Musée Océanographique de Monaco, 1923, 11.

29. Portier, *Les Symbiotes,* v–vi.

30. Paul Portier, "Recherche physiologiques sur les insectes aquatiques," Thèse Faculté des Sciences, University of Paris, 1911.

31. E. Duclaux, "Sur la germination dans un sol riche en matières organiques, mais exempt de microbes," *Comptes rendus de l'Académie des Sciences* 100 (1885): 66–68, and Louis Pasteur, "Observations relatives à la Note précédente de M. Duclaux," 68.

32. See Thomas D. Luckey, *Germfree Life and Gnotobiology* (New York: Academic Press, 1963), 61.

33. For a historical account of the theory of phagocytosis, see Tauber and Chernyak, *Metchnikoff and the Origins of Immunology.*

34. Paul Portier, "Résistance aux agents chemiques de certaines races du *B. subtilis* provenant des Insectes," *Comptes rendus de l'Académie des Sciences* 161 (1915): 397–402; "Sur la présence de microcoques dans le snag des typhoïdiques provenant du front," *Comptes rendus de la Société de Biologie* 78 (1915): 440–45.

35. Raphaël Dubois, "Symbiotes, Vacuolides, Mitochondries et Leucites," *Comptes rendus de la Société de Biologie* 82 (1919): 473–75; "Les Vacuolides Sont-Elles des Symbiotes?" *Comptes rendus de la Société de Biologie* 82 (1919): 475–77.

36. For biographical information on Umberto Pierantoni, see M. Salfi, "Umberto Pierantoni, 1876–1959," *Bollettino di Zoologia* 27 (1960): 21–30; Riccardo Pierantoni, "Studiosi e Naturalisti del Nostro Tempo: Umberto Pierantoni," *Bollettino Sezione Campana ANISN* 4 (1992): 11–19. For his early work on symbiosis, see, for example, Umberto Pierantoni, "Su alcuni Euplotidae del golfo di Napoli," *Bollettino della Società dei Naturalisti in Napoli* 23 (1909): 53–62; "Struttura ed evoluzione dell'organo simbiotico di *Pseudococcus citri* Risso, e ciclo biologico del *Coccidomyces dactylopii* Buchner," *Archiv für Pristenkunde* 31 (1913): 300–316; "Part II. Origine ed evoluzione degli organi sessuali maschili. Ermafroditismo," *Archivio Zoologicao Italiano* 7 (1914): 27–46; "Sulla biofotogenesi simbiotica," *Bollettino della Società die Naturalista in Napoli* 34 (1922): 307–9; "La fosforescenza e la simbiosi in *Micro-scolex phosphoreus,*" *Bollettino della Società die Naturalista in Napoli* 36 (1924): 179–95.

37. Paul Buchner, *Endosymbiosis of Animals with Plant Microorganisms* New York: Interscience, 1965), 73.

38. Quoted in Maurice Caullery, *Parasatism and Symbiosis* (London: Sidgwick and Jackson, 1950), 279.

39. Pierre Mazé, "Les microbes des nodosités des Légumineuse," *Annale de l'Institut Pasteur,* vols. 11, 12, 13 (1897).

40. Noël Bernard, "Etudes sur la Tubérisation," *Revue Générale de Botanique* 14 (1902): 5–25; "Remarques sur l'immunité chez les plantes," *Bulletin de l'Institut Pasteur* 7 (1909): 369–86.

41. Gabriel Bertrand, "Sur une ancienne expérience de M. Berthelot relative à la transformation de la glycérine en sucre par le tissue testiculaire," *Comptes rendus de l'Académie des Sciences* 133 (1901): 887–92.

42. V. Galippe, *Parasitisme normal et microbiose* (Paris: Masson, 1917).

43. Pierre Béchamp, *Microzymas* (Montpellier, 1875).

44. Paul Portier, "Recherches sur les microorganismes symbiotiques dans la série animale," *Comptes rendus des séances de l'Académie des Sciences* 165 (1917): 196–99.

45. Ibid., 197.

46. Paul Portier, "Rôle Physiologique des Symbiotes," *Comptes rendus des séances de l'Académie des Sciences* 165 (1917): 267–69, 268.

47. Ibid., 269.

48. H. Bierry and Paul Portier, "Action des symbiotes sur les Constituants des Graisses," *Comptes rendus de l'Académie des Sciences* 166 (1918): 1055–57.

49. H. Bierry and Paul Portier, "Innocuité de l'Introduction des Symbiotes dans le Milieu Intérieur des Vertébrés," *Comptes rendus de la Société de Biologie* 81 (1918): 480–81.

50. Henri Bierry and Paul Portier, "Vitamines et Symbiotes," *Comptes rendus de l'Académie des Sciences* 166 (1918): 963–67.

51. Paul Portier to Jules Richard, 17 December 1917, Archives du Musée Océanographique de Monaco.

52. Paul Portier to Jules Richard, 4 September 1918, Archives du Musée Océanographique de Monaco.

53. Portier, *Les Symbiotes,* viii.

54. Ibid., 19.

55. Ibid., xiv.

56. Ibid., ix.
57. Ibid., 82–83.
58. Ibid., xii.
59. Ibid., 23.
60. Ibid., 68.
61. Ibid., 75.
62. Ibid., 76.
63. Alexandre Guilliermond, "Etat actuel de la question de l'évolution et du rôle physiologique des mitochondries d'après les travaux récents de cytologie végétale," *Revue Générale Botanique* 26 (1914): 129–43, quoted in Portier, *Les Symbiotes,* 76.
64. Portier, *Les Symbiotes,* 77.
65. Ibid.
66. Ibid., 78–79.
67. Ibid., 79.
68. For an account of the development of studies of nitrogen fixation, see P. W. Wilson and E. B. Fred, "The Growth of a Scientific Literature," *Scientific Monthly* 41 (1935): 240–50.
69. Portier, *Les Symbiotes,* 80–81.
70. Ibid., 81.
71. Ibid., 93.
72. Ibid., 100.
73. Ibid.
74. Ibid., 101.
75. Ibid., 102.
76. Ibid., 105–6.
77. Ibid., 106.
78. Ibid., 131.
79. Ibid., 120.
80. Ibid., 149.
81. Ibid., 151.
82. Ibid., 157.
83. Ibid., 294.

Chapter 7

1. Anna Drzewina, review of *Les Symbiotes, Revue Scientifique* 57 (1919): 414–15, 414.
2. Paul Portier to Prince Albert I de Monaco, 28 October 1918, Archives du Palais Princier de Monaco, C. 800.
3. Victor Galippe, *Parasitism Normal et Microbiosis* (Paris: Masson, 1917).
4. Yves Delage and Marie Goldsmith, comments on "Les Grands problèmes de la biologie générale," *L'Année Biologique* 22 (1917): xii.
5. Yves Delage, comments following Bierry and Portier, "Les Vitamines et les symbiotes," *Comptes rendus de la Société de Biologie* 166 (1918): 966.
6. Paul Portier, note following P. Masson and C. Regaud, "Apparition et Pullulation des microbes dans le tissu lymphoide de l'appendice caecal du lapin, au cours du development," *Comptes rendus de la Société de Biologie* 71 (1919): 32.
7. Paul Portier to Prince Albert I de Monaco, 30 December 1918, Archives du Musée Océanographique de Monaco.

8. Paul Portier to Jules Richard, 21 January 1919, Archives du Musée Océanographique de Monaco.

9. Yves Delage and Marie Goldsmith, review of *Les Symbiotes, L'Année Biologique* 23 (1919): 313–17, 315, 316.

10. Ibid., 316.

11. Jules Richard to Prince Albert I de Monaco, 24 January 1919, Archives du Musée Océanographique de Monaco.

12. Paul Portier to Jules Richard, 2 February 1919, Archives du Musée Océanographique de Monaco.

13. Louis Martin, comments, *Comptes rendus de la Société de Biologie 82 (1919): 128–29, 128.*

14. Ibid., 129.

15. Etienne Marchoux, comments, *Comptes rendus de la Société de Biologie* 82 (1919): 129–30, 130.

16. Claude Regaud, comments, *Comptes rendus de la Société de Biologie* 82 (1919), 131.

17. Quoted in Henri Bierry, comments, *Comptes rendus de la Société de Biologie* 82 (1919): 131–32, 131.

18. Paul Portier, comments, *Comptes rendus de la Société de Biologie* 82 (1919): 132.

19. Société de Biologie, statement regarding debate over *Les Symbiotes, Comptes rendus de la Société de Biologie* 82 (1919): 133.

20. Claude Regaud, "Mitochondrie et symbiotes," *Comptes rendus de la Société de Biologie* 82 (1919): 244–47, 246.

21. Ibid., 247.

22. Paul Portier, comments, *Comptes rendus de la Société de Biologie* 82 (1919): 247–50.

23. Ibid., 249–50.

24. Claude Regaud, comments, *Comptes rendus de la Société de Biologie* 82 (1919): 250–51.

25. Ibid., 251.

26. Henri Bierry, "Avitaminose et carence," *Comptes rendus de la Société de Biologie* 82 (1919): 307–9.

27. Alexandre Guilliermond, "Mitochondrie et symbiotes," *Comptes rendus de la Société de Biologie* 82 (1919): 309–12.

28. Ibid., 310.

29. Ibid., 312.

30. Ibid.

31. E. Laguesse, "Mitochondries et symbiotes," *Comptes rendus de la Société de Biologie* 82 (1919): 337–39.

32. Raphaël Dubois, "Symbiotes, vacuolides, mitochondries et leucites," *Comptes rendus de la Société de Biologie* 82 (1919): 473–75.

33. Ibid., 474.

34. Ibid., 476.

35. Ibid., 474.

36. Ibid., 475.

37. Portier, *Les Symbiotes,* 70.

38. Raphaël Dubois, "Les Vacuolides sont-elles des symbiotes?" *Comptes rendus de la Société de Biologie* 82 (1919): 475–77.

39. L. Matruchot, review of *Les Symbiotes, Revue générale des sciences pure et appliquées* 30 (1919): 535–36.

40. Paul Portier to Jules Richard, 1 April 1919, Archives du Musée Océanographique de Monaco.
41. Ibid.
42. Paul Portier to Prince Albert I de Monaco, 13 May 1919, Archives du Musée Océanographique de Monaco.
43. Jules Richard to Prince Albert I de Monaco, 22 May 1919, Archives du Musée Océanographique de Monaco.
44. Paul Portier, *Exposé des titres et travaux Scientifiques* (Paris: Jouve, 1919), 42.
45. Paul Portier to Jules Richard, 10 October 1919, Archives du Musée Océanographique de Monaco.
46. Bernard Chardere, Guy Borgé, and Marjorie Borgé, *Les Lumières* (Payot Lausanne: Bibliothèque des Arts Paris, 1986). Léon Binet, "Notice Nécrologique sur Auguste Lumière," *Comptes rendus de l'Académie des Sciences* 258 (1954): 1944–45.
47. Auguste Lumière, *Principaux travaux scientifiques* (Lyon: Leon Sézanne, 1935).
48. Auguste Lumière, *Les Mythes des symbiotes* (Paris: Masson, 1919), xi.
49. Ibid., viii.
50. Ibid., 20.
51. Ibid., 28–29.
52. Ibid., 31.
53. Ibid., 32.
54. Ibid., 33–34.
55. Ibid., 39.
56. Ibid., 75.
57. Ibid., 72.
58. A. Drzewina, "Review of *Les Symbiotes,*" *Revue scientifique* 57 (1919): 414–415.
59. Lumière, *Les Mythes des symbiotes,* 155–56.
60. Paul Portier to Jules Richard, 5 November 1919, Archives du Musée Océanographique de Monaco.
61. Paul Portier to Jules Richard, 11 November 1919, Archives du Musée Océanographique de Monaco.
62. Paul Portier to Prince Albert I de Monaco, 5 May 1920, Archives du Palais Princier de Monaco, C. 805.
63. H. Bierry, E. Marchoux, L. Martin, and P. Portier, "Rapport de la commission nommé par la Société de Biologie dans la séance du 8 Février, 1919," *Comptes rendus de la Société de Biologie* 83 (1920): 654.
64. Paul Portier, "Rapport sur les travaux du laboratoire de physiologie," *Rapports des differents services de l'Institut de Paris et du Musée de Monaco* (1922) p. 7, Archives du Musée Océanographique de Monaco.
65. Maurice Caullery, *Le Parasitisme et symbiose* (Paris: Librairie Octave Doin, 1922).
66. Ibid., 7. See also Maurice Caullery, *Parasitism and Symbiosis,* trans. Averil M. Lysaght (London: Sidgwick and Jackson, 1952), xii.
67. A detailed biography of d'Herelle is being prepared by William C. Summers.
68. Paul Portier to Jules Richard, 6 July 1923, Archives du Musée Océanographique de Monaco.
69. For an account of this priority dispute during this period see Paul Hauduroy, *Le Bactériophage de d'Hérelle* (Paris: Librarie Le François, 1925), 1–6.
70. D'Herelle believed that the phenomenon reported by Twort was not due to a

bacteriophage. See Félix d'Herelle, *The Bacteriophage. Its Rôle in Immunity,* trans. George H. Smith (Baltimore: Williams and Wilkins, 1922), 17.

71. Ibid., 149.
72. Ibid., 79.
73. Ibid., 273–74.
74. Ibid.
75. Félix d'Herelle, *The Bacteriophage and Its Behavior,* trans. George H. Smith (Baltimore: William and Wilkins, 1926), 211.
76. Ibid., 319.
77. Ibid., 320.
78. Ibid., 320.
79. Ibid., 212.
80. Ibid., 320.
81. Félix d'Herelle, "Bacterial Mutations," *Yale Journal of Biology and Medicine* 4 (1931): 55–57, 56.
82. Paul Portier, *Notice sur les titres et travaux scientifiques de Paul Portier* (Paris: A. Maretheux et L. Pactat, 1936), 163, 164.
83. K. Mansour and J. J. Mansour-Bek, "The Digestion of Wood by Insects and the Supposed Role of Micro-Organisms," *Biological Reviews* 6 (1934): 363–82, 364.
84. Portier, *Notice sur les titre et travaux,* 169.

Chapter 8

1. I. E. Wallin, *Symbionticism and the Origin of Species* (Baltimore: Williams and Wilkins, 1927), 8.
2. Anton Koch, "Dedication to Paul Buchner," in S. Mark Henry, ed., *Symbiosis,* vol. 2 (New York: Academic Press, 1967), v–viii.
3. Paul Buchner, *Endosymbiosis of Animals with Plant Microorganisms,* trans. Bertha Mueller (New York: Interscience, 1965), 29.
4. Paul Buchner, *Tier und Pflanze in intrazellularer Symbiose* (Berlin: Borntraeger, 1921).
5. Ibid., 74.
6. Quoted in Wallin, *Symbionticism,* 146.
7. George H. F. Nuttall, "Symbiosis in Animals and Plants," *Report of the British Association for the Advancement of Science* (1923): 197–214, 197.
8. Ibid., 214.
9. K. F. Meyer, "The 'Bacterial Symbiosis' in the Concretion Deposits of Certain Operculate Land Mollusks of the Families Cyclostomatidae and Annularidea," *Journal of Infectious Diseases* 36 (1925): 1–107, 96.
10. Ibid., 97–98.
11. L. R. Cleveland, "Symbiosis among Animals with Special Reference to Termites and Their Intestinal Flagellates," *Quarterly Review of Biology* 1 (1926): 51–60; 54.
12. Ibid.
13. Ibid., 55.
14. Ibid., 53.
15. See Donna C. Mehos, "The Defeat of Ivan E. Wallin and His Symbiotic Theory of Evolution" unpublished manuscript, 1990. I. E. Wallin, "On the Brancial Epithelium of Ammocoetes," *Anatomical Record* 14 (1918): 205–15.
16. E. V. Cowdry, "The Development of the Cytoplasmic Constituents of the

Nerve Cells of the Chick," *American Journal of Anatomy* 15 (1914): 389–429; "The General Significance of Mitochondria," *American Journal of Anatomy* 19 (1916): 423–46; "The Mitochondrial Constituents of Protoplasm," *Carnegie Institution Publications, Contributions to Embryology* 8, no. 25 (1918): 41–144.

17. Cowdry, "The Mitochondrial Constituents," 96.

18. Ibid.

19. T. H. Morgan, "Genetics and the Physiology of Development," *American Naturalist* 60 (1926): 489–515, 496.

20. Cowdry, "The Mitochondrial Constituents," 99.

21. Anonymous, review of Portier's *Les Symbiotes, Nature* 103 (1919): 483.

22. I. E. Wallin, "On the Nature of Mitochondria. I. Observations on Mitochondria Stating Methods Applied to Bacteria. II. Reactions of Bacteria to Chemical Treatment," *American Journal of Anatomy* 30 (1920: 203–29, 225.

23. Ibid., 225.

24. Ivan E. Wallin, "On the Nature of Mitochondria. V. A Critical Analysis of Portier's 'Les Symbiotes,' " *Anatomical Record* 25 (1923): 1–5, 1–2.

25. Ibid., 2.

26. Ibid.

27. Ibid., 5.

28. Ibid., 2.

29. Ibid., 4.

30. Ibid.

31. Ibid.

32. Ivan E. Wallin, "On The Nature of Mitochondria. VII. The Independent Growth of Mitochondria in Culture Media," *American Journal of Anatomy* 33 (1924): 147–73, 147.

33. Wallin, "On the Nature of Mitochondria V," 3.

34. E. V. Cowdry and P. K. Olitsky, "Differences between Mitochondria and Bacteria, *Journal of Experimental Medicine* 36 (1922): 521–36, 521.

35. Ibid., 521.

36. Ibid., 522–23.

37. Ibid.

38. W. H. Park and A. W. Williams, *Pathogenic Microörganisms* (Philadelphia, 1920), 24.

39. "Are Mitochondria Identical with Bacteria?" *Journal of the American Medical Association* 79 (1922): 1848–49, 1849.

40. Ivan E. Wallin, "The Mitochondria Problem," *American Naturalist* 57 (1923): 255–61, 261.

41. Ibid., 260.

42. Ibid., 259.

43. Ibid., 257.

44. Wallin, *Symbionticism,* 40.

45. Wallin, "The Mitochondria Problem," 261.

46. Wallin, "On the Nature of Mitochondria. VII," 147.

47. Ibid., 159.

48. Ibid., 167.

49. E. B. Wilson, *The Physical Basis of Life* (New Haven: Yale University Press, 1923), 28–29.

50. John F. Fulton, "Animal Chlorophyll: Its Relation to Haemoglobin and to Other Animal Pigments," *Quarterly Journal of Microscopical Science* 66 (1922): 340–96.

51. E. B. Wilson, *The Cell in Development and Heredity* (New York: Macmillan, 1925), 738.

52. Ibid., 739.

53. Ivan E. Wallin, "On the Nature of Mitochondria. IX. Demonstration of the Bacterial Nature of Mitochondria," *American Journal of Anatomy* 36 (1925): 131–46, 140.

54. Ibid., 139.

55. Ivan E. Wallin, "On the Nature of Mitochondria. III. The Demonstration of Mitochondria by Bacterial Methods. IV. A Comparative Study of the Morphogenesis of Root-Nodule Bacteria and Chloroplasts," *American Journal of Anatomy* (1922): 451–73, 463–64.

56. Ibid., 464.

57. Ibid.

58. Ibid., 465.

59. Ivan E. Wallin, "Symbionticism and Prototaxis, Two Fundamental Biological Principles," *Anatomical Records* 26 (1923): 65–73, 67.

60. Ibid., 71.

61. Wallin, *Symbionticism and the Origin of Species,* 67, 63.

62. Wallin, "On the Nature of Mitochondria. III," 461.

63. Wallin, "On the Nature of Mitochondria. IX," 141.

64. Wallin, *Symbionticism,* 96.

65. Ibid., 114.

66. Ibid., 63.

67. Ibid., 147.

68. See Garland E. Allen, "Hugo de Vries and the Reception of the 'Mutation Theory,' " *Journal of the History of Biology* 2 (1968): 55–87. Peter Bowler, *The Eclipse of Darwinism: AntiDarwinian Evolution Theories in the Decades around 1900* (Baltimore: Johns Hopkins University Press, 1983).

69. Wallin, *Symbionticism,* 137.

70. Ibid., 6.

71. Ibid., 134.

72. Ibid., 134–35.

73. Unlike Merezhkovskii, however, Wallin harmonized the phylogenetic origin of chloroplasts from blue-green algae with the prevalent belief they were derived in ontogeny from mitochondria. He simply suggested that the blue-green algae had been modified upon acquisition of the symbiotic relationship. Ibid., 109–11.

74. Ibid., 142.

75. Wallin, "On the Nature of Mitochondria. IX," 142.

76. Wallin, *Symbionticism,* 126–27.

77. R. G. Swinburne, "The Presence and Absence Theory," *Annals of Science* 18 (1962): 131–46. See also discussions by Jan Sapp in *Where the Truth Lies* (New York: Cambridge University Press, 1990).

78. Wallin, *Symbionticism,* 120–21.

79. Ibid., 121.

80. Ibid., 127.

81. William Bateson, *Problems of Genetics* (New Haven: Yale University Press, 1913), 88.

82. Ibid., 90.

83. Wallin, *Symbionticism,* 143.

84. See, for example, K. David Patterson, *Pandemic Influenza, 1700–1900. A Study in Historical Epidemiology* (Totowa, N.J.: Rowman and Littlefield, 1986).

85. Wallin, *Symbionticism,* 143.
86. Ibid., 144–45.
87. J. Brontë Gatenby, "Nature of Cytoplasmic Inclusions," *Nature* 121 (1928): 164–65, 165.
88. Paul de Kruif, *Microbe Hunters* (New York: Harcourt, Brace, 1926), 334.
89. See H. J. Muller, "The Gene," *Proceedings of the Royal Society of London B,* 134 (1947): 1–37.
90. H. J. Muller, "The Gene as the Basis of Life," *Proceedings of the International Congress of Plant Sciences* 1 (1929): 897–921.
91. For discussions of disciplinary and national contributions to the "evolutionary synthesis" see Ernst Mayr and William B. Provine, eds., *The Evolutionary Synthesis: Perspectives on the Unification of Biology* (Cambridge, Mass.: Harvard University Press, 1980).
92. See, for example, C. M. Child, *Senescence and Rejuvenescence* (Chicago: University of Chicago Press, 1915), 11–12.
93. E. G. Conklin, "Cell and Protoplasm concepts: Historical Account," in G. R. Moulton, ed., *The Cell and Protoplasm* (Washington: Science Press, 1940), 6–19, 17.
94. Ibid., 17.
95. Conklin, "Cell and Protoplasm Concepts," 18.
96. See Jonathan Harwood, "The Reception of Morgan's Chromosome Theory in Germany: Interwar Debate over Cytoplasmic Inheritance," *Medizinhistorisches Journal* 19 (1984): 3–32; "Geneticists and the Evolutionary Synthesis in Interwar Germany," *Annals of Science* 42 (1985): 279–301. Jonathan Harwood, *Styles of Scientific Thought* (Chicago: University of Chicago Press, 1992). Jan Sapp, *Beyond the Gene* (New York: Oxford University Press, 1987).
97. Sapp, *Beyond the Gene,* 80–86.
98. E. M. East, "The Nucleus-Plasma Problem," *American Naturalist* 68 (1934): 402–39, 409–10.
99. F. G. Gregory, "A Discussion on Symbiosis Involving Micro-organisms," *Proceedings of the Royal Society of London B* 139 (1951): 202–3, 202.

Chapter 9

1. Maurice Caullery, *Parasitism and Symbiosis,* trans. Averil M. Lysaght (London: Sidgwick and Jackson, 1952), 2.
2. Marshall Hertig, W. H. Taliaferro, and Benjamin Schwartz, "Supplement to the Report of the Twelfth Annual Meeting of the American Society of Parasitologists," *Journal of Parasitology* 23 (1937): 325–29, 327.
3. Ibid., 328.
4. George H. Nuttall, "Symbiosis in Animals and Plants," *Report of the British Association for the Advancement of Science* (1923): 197–214, 198.
5. K. F. Meyer, "The 'Bacterial Symbiosis' in the Concretion Deposits of Certain Operculate Land Mollusks of the Families Cyclostomatidae and Annularidae," *Journal of Infectious Diseases* 36 (1925): 1–107, 94–95.
6. Ibid., 95.
7. Ibid., 96.
8. Ibid.
9. Ibid., 97.
10. Ibid.

11. Ibid.

12. L. R. Cleveland, "Symbiosis among Animals with Special Reference to Termites and Their Intestinal Flagellates," *Quarterly Review of Biology* 1 (1926): 51–56.

13. Ibid., 51.

14. Ibid., 53.

15. Ivan E. Wallin, *Symbionticism and the Origin of Species* (Baltimore: Williams and Wilkins, 1927), 65.

16. Ibid., 66.

17. Ibid., 59.

18. H. G. Wells, Julian Huxley, and G. P. Wells, *The Science of Life,* vol. 3 (New York: Doubleday, Doran and Co., 1930), 992.

19. Ibid., 930.

20. Ibid.

21. Ibid., 931–32.

22. Ibid., 932.

23. Ibid., 935.

24. Ibid., 936.

25. Donald Worster, *Nature's Economy. A History of Ecological Ideas* (Cambridge: Cambridge University Press, 1977), 328–31. Gregg Mitman, *The State of Nature. Ecology, Community and American Social Thought* (Chicago: University of Chicago Press, 1992).

26. W. C. Allee, "Where Angels Fear to Tread: A Contribution from General Sociology to Human Ethics," *Science* 97 (1943): 517–25. Reprinted in Arthur L. Caplan, *The Sociobiology Debate. Readings on Ethical and Scientific Issues* (New York: Harper and Row, 1978), 41–56, 51. All citations refer to reprint.

27. Ibid., 52.

28. Ibid., 55.

29. Ibid., 55.

30. Ibid., 55.

31. Ibid., 56.

32. W. C. Allee, Alfred E. Emerson, Orlando Park, Thomas Park, and Karl P. Schmidt, *Principles of Animal Ecology* (Philadelphia: W. B. Saunders, 1949), 694.

33. Ibid., 32.

34. Ibid., 245–53.

35. Ibid., 254.

36. Ibid., 710.

37. Ibid., 728.

38. Ibid., 712.

39. Ibid., 711.

40. Ibid., 713.

41. Ibid., 712.

42. Ibid., 717–18.

43. Ibid., 716.

44. Ibid.

45. Ibid., 718.

46. Ibid.

47. Ibid.

48. For a discussion of this transition see Mitman, *The State of Nature.*

49. Eugene P. Odum, *Fundamentals of Ecology* (Philadelphia: W. B. Saunders, 1959), 228.

50. Ibid., 242.
51. Ibid., 243–44.
52. Caullery, *Parasitism and Symbiosis,* xi.
53. Ibid., 217.
54. Ibid., 221.
55. Ibid., 217.
56. Ibid., 218.
57. Allee et al., *Principles of Animal Ecology,* 713–14.
58. Caullery, *Parasitism and Symbiosis,* 220.
59. Ibid., 259.
60. Noël Bernard, "Remarques sur l'immunité chez les plantes," *Bulletin de l'Institut Pasteur* 7 (1909): 369–86.
61. Caullery, *Parasitism and Symbiosis,* 276.
62. H. G. Thornton, "Introduction. The Symbiosis between *Rhizobium* and Leguminous Plants and the Influence on This of the Bacterial Strain," *Proceedings of the Royal Society of London B* 139 (1951): 171–85, 171.
63. F. G. Gregory, "A Discussion on Symbiosis Involving Micro-organisms—General Discussion," *Proceedings of the Royal Society of London B* 139 (1951): 202–3, 202.
64. Ibid.
65. F. Baker, ibid., 171.
66. Ibid.
67. Ibid.
68. Paul Fildes, ibid., 205.
69. S. W. Orenski, "Intermicrobial Symbiosis," in S. Mark Henry ed., *Symbiosis,* vol. 1 (New York: Academic Press, 1966), 1–35, 1.
70. Caullery, *Parasitism and Symbiosis,* 277.
71. Ibid., 285.
72. Wells, Huxley, and Wells, *The Science of Life,* 2: 505.
73. Ibid., 507.
74. Ibid., 507.

Chapter 10

1. Joshua Lederberg, "Cell Genetics and Hereditary Symbiosis," *Physiological Reviews* 32 (1952): 403–30, 403–4.
2. See, for example, Horace Freeland Judson, *The Eighth Day of Creation. The Makers of the Revolution in Biology* (New York: Simon and Schuster, 1979). Lily E. Kay, *The Molecular Vision of Life: Caltech, The Rockefeller Foundation, and the Rise of the New Biology* (New York: Oxford University Press, 1993). Robert Olby, *The Path to the Double Helix* (Seattle: University of Washington Press, 1974). Jan Sapp, *Where the Truth Lies. Franz Moewus and the Origins of Molecular Biology* (New York: Cambridge University Press, 1990).
3. D. Joravsky, *The Lysenko Affair* (Cambridge, Mass.: Harvard University Press, 1970). Richard Lewontin and Richard Levins, "The Problem of Lysenkoism," in Hilary Rose and Steven Rose, eds., *The Radicalization of Science* (London: Macmillan, 1976), 32–65. Z. A. Medvedev, *The Rise and Fall of T. D. Lysenko,* trans. I. Michael Lerner (New York: Anchor Books, 1971).
4. Jan Sapp, *Beyond the Gene. Cytoplasmic Inheritance and the Struggle for Au-*

thority in Genetics (New York: Oxford University Press, 1987); "Concepts of Organization: The Leverage of Ciliate Protozoa," in Scott Gilbert, ed., *A Conceptual History of Modern Developmental Biology* (New York: Plenum, 1991), 229–58.

5. See Sapp, *Beyond the Gene.*

6. Boris Ephrussi, "Remarks on Cell Heredity," in L. C. Dunn, ed., *Genetics in the Twentieth Century* (New York: Macmillan, 1951), 241–62.

7. Boris Ephrussi, *Nucleo-Cytoplasmic Relations in Microorganisms* (Oxford: Clarendon Press, 1953).

8. André Lwoff, *Problems of Morphogenesis in Ciliates. The Kinetosomes in Development, Reproduction and Evolution* (New York: Wiley, 1951), 15.

9. T. M. Sonneborn, "The Cytoplasm in Heredity," *Heredity* 4 (1950): 11–36.

10. Edgar Altenburg, "The Viroid theory in Relation to Plasmagenes, Viruses, Cancer and Plastids," *American Naturalist* 80 (1946): 559–67; "The Symbiont Theory in Explanation of the Apparent Cytoplasmic Inheritance in Paramecium," *American Naturalist* 80 (146): 661–62.

11. H. J. Muller, "Variation Due to Change in the Individual Gene," *American Naturalist* 56 (1922): 32–50, 48–49.

12. Gunther S. Stent, *Molecular Biology of Bacterial Viruses* (San Francisco: W. H. Freeman, 1963), 19.

13. For a discussion of the concept of virus, see André Lwoff, "Lysogeny," *Bacteriological Reviews* 17 (1953): 270–337; "The Concept of Virus," *Journal of General Microbiology* 17 (1957): 240–53.

14. Ellen G. Strauss, James H. Strauss, and Arnold J. Levine, "Virus Evolution," in B. N. Fields, D. M. Knipe, et al., eds., *Virology* (New York: Raven Press, 1990), 167–90.

15. Altenburg, "The Viroid Theory," 559.

16. Ibid.

17. Ibid., 563.

18. Ibid., 565–566.

19. C. D. Darlington, "Heredity, Development and Infection," *Nature* 154 (1944): 164–69.

20. Altenburg, "The Symbiont Theory," 662.

21. Philippe l'Héritier, "Sensitivity to CO_2 in *Drosophila*—A Review," *Heredity* 2 (1948): 325–48, 346.

22. H. J. Muller, "The Development of the Gene Theory," in L. C. Dunn, ed., *Genetics in the Twentieth Century* (New York: Macmillan, 1951), 77–100, 82–83.

23. Ibid., 83.

24. See Jan Sapp, "Inside the Cell: Genetic Methodology and the Case of the Cytoplasm," in J. Schuster and R. Yeo, eds., *The Politics and Rhetoric of Scientific Method* (Dordrecht: Reidel, 1986), 167–202.

25. Sonneborn, "The Cytoplasm in Heredity," 22.

26. See Gregg Mitman, *Cooperative Nature: Animal Ecology and Community at Chicago, 1900–1950* (Chicago: University of Chicago Press, 1992).

27. Theodosius Dobzhansky, *Genetics and the Origin of Species* (New York: Columbia University Press, 1951), 285.

28. Ibid., 286.

29. Ibid., 230.

30. C. D. Darlington, "Mendel and the Determinants," in L. C. Dunn, ed., *Genetics in the Twentieth Century* (New York: Macmillan, 1951), 315–22, 320.

31. Ibid., 320.

32. Ibid., 327.
33. Ibid., 331.
34. See Susan Wright, "Recombinant DNA Technology and Its Social Transformation, 1972–1982," *Osiris* 2 (1986): 303–60, 310.
35. Joshua Lederberg, "Bacterial Variation," *Annual Review of Microbiology* 3 (1949): 1–22, 18.
36. See Joshua Lederberg, "Genetic Transduction," *American Scientist* 44 (1956): 264–80, 271.
37. E. M. Lederberg and Joshua Lederberg, "Genetic Studies of Lysogeny in *Escherichia colí,*" *Genetics* 38 (1953): 51–64, 61.
38. Stent, *Molecular Biology of Bacterial Viruses,* 14.
39. Ibid.
40. Quoted in Lwoff, "Lysogeny," 277.
41. Charles Galperin, "Le bactériophage, la lysogenie et son déterminisme génétique," *History and Philosophy of the Life Sciences* 9 (1987): 175–24.
42. Judson, *The Eighth Day of Creation,* 373.
43. Joshua Lederberg, "Genetic Studies in Bacteria," in L. C. Dunn, ed., *Genetics in the Twentieth Century* (New York: Macmillan, 1951), 263–90, 286.
44. Ibid., 286.
45. Ibid.
46. Ibid., 287.
47. Lederberg, "Cell Genetics and Hereditary Symbiosis."
48. Contrary ideas can still be found in the modern literature. See, for example, D. Ross Williams and Christopher M. Thomas, "Active Partitioning of Bacterial Plasmids," *Journal of General Microbiology* 138 (1920: 1–16, 1. The authors assert: "Bacterial plasmids, by definition are not essential for host cell survival."
49. Lederberg, "Cell Genetics and Hereditary Symbiosis," 403–4.
50. Ibid., 413.
51. Ibid., 419.
52. Ibid., 415.
53. Ibid., 421.
54. Ibid., 422–23.
55. Ibid.
56. Ibid., 424.
57. Ibid.
58. Ibid., 425.
59. Ibid., 425–26.
60. Philippe l'Héritier, "Les Virus Integrés et l'Unité Cellulaire," *L'Anné Biologique* 31 (1955): 481–96.
61. T. M. Sonneborn, "Heredity, Development and Evolution in *Paramecium,*" *Science* 175 (1955): 1100–1102.
62. See Lederberg, "Genetic Transduction," 275.
63. Ephrussi, *Nucleo-Cytoplasmic Relations in Microorganisms,* 4.
64. See D. L. Nanney, "Epigenetic Control Systems," *Proceedings of the National Academy of Sciences* 44 (1958): 327–35. Boris Ephrussi, "The Cytoplasm and Somatic Cell Variation," *Journal of Cellular and Comparative Physiology* 52 (1958): 35–53.
65. T. M. Sonneborn, "The Differentiation of Cells," *Proceeding of the National Academy of Sciences* 51 (1964): 915–29, 918.
66. François Jacob and Jacques Monod, "Elements of Regulatory Circuits in Bacte-

ria," in R. J. C. Harris, ed., *Biological Organization at the Cellular and Subcellular Level* (London: Academic Press), 1–24.

67. For a historical account of the controversy over the inheritance of supramolecular properties of the cell, see Sapp, "Concepts of Organization."

68. See, for example, Winston Salser, "Non-genetic Biological Information Mechanisms," *Perspectives in Biology and Medicine* 4 (1961): 177–98.

Chapter 11

1. P. S. Nutman and Barbara Mosse, "Editors Preface," in P. S. Nutman and Barbara Mosse, eds., *Symbiotic Associations. Thirteenth Symposium of the Society for General Microbiology* (Cambridge: Cambridge University Press, 1963), ix–x, x.

2. Joshua Lederberg to Walter Carter, 22 June 1954, Lederberg Papers, Rockefeller University.

3. Richard Goldschmidt to Joshua Lederbert, 25 April 1953, Lederberg Papers, Rockefeller University.

4. Richard Goldschmidt to Joshua Lederberg, 21 March 1954, Lederberg Papers, Rockefeller University.

5. Paul Buchner, *Endosymbiosis of Animals with Plant Microorganisms,* rev. Engl. ed. (New York: Interscience, 1965).

6. Clark P. Read, "A Science of Symbiosis," *Bulletin of the American Institute for Biological Sciences* 7 (1958): 16–17; 17.

7. Ibid., 17.

8. Jane Thomas Rowland, "Symbiosis: Rich, Exciting, Neglected Topic," *American Biology Teacher* 36 (1974): 77–82.

9. See, for example, M. Burton, *Animal Partnerships* (New York: Frederick Warne, 1969). R. H. Dudley, *Partners in Nature* (New York: Funk and Wagnalls, 1965). P. Shuttlesworth, *Natural Partnerships: The Story of Symbiosis* (New York: Doubleday, 1969); *Togetherness in the World of Nature: Symbiosis,* in Audubon Nature Bulletin Set NB 7 on Ecology (New York: National Audubon Society, 1971). H. Simon, *Partners, Guests, and Parasites: Coexistence in Nature* (New York: Viking Press, 1962).

10. See discussion in S. Mark Henry, ed., *Symbiosis,* vol. 1 (New York: Academic Press, 1966), xiii.

11. Nutman and Mosse, "Editors Preface," x.

12. Ibid.

13. Peter Baldry, *The Battle against Bacteria. A Fresh Look* (Cambridge: Cambridge University Press, 1965).

14. René Dubos and Alex Kessler, "Integrative and Disintegrative Factors in Symbiotic Associations," in P. B. Nutman and Barbara Mosse, eds., *Symbiotic Associations. Thirteenth Symposium of the Society for General Microbiology* (Cambridge: Cambridge University Press, 1963), 1–11, 4.

15. René J. Dubos, *The Bacterial Cell* (Cambridge, Mass.: Harvard University Press, 1945).

16. René J. Dubos, "Integrative and Creative Aspects of Infection," in M. Pollard, ed., *Perspectives in Virology,* vol. 2 (Minneapolis: Burgess, 1961), 200–205, 200.

17. Ibid., 204.

18. Ibid., 204.

19. Ibid., 10.

20. M. P. Droop, "Algae and Invertebrates in Symbiosis," in P. S. Nutman and Barbara Mosse, eds., *Symbiotic Associations. Thirteenth Symposium of the Society for General Microbiology* (Cambridge: Cambridge University Press, 1963), 171–99.

21. Stephen J. Karakashian, "Growth of *Paramecium bursaria* as Influenced by the Presence of Algal Symbionts," *Physiological Zoology* 36 (1963): 52–68, 66.

22. S. J. Karakashian and R. W. Siegel, "A Genetic Approach to Endocellular Symbiosis," *Experimental Parasitology* 17 (1965): 103–22, 103.

23. Ibid., 118.

24. See R. Lang, "Bacterial Symbiosis with Plants," in S. M. Henry, ed., *Symbiosis,* vol. 1 (New York: Academic Press, 1966), 99–170.

25. Paul Buchner, *Endosymbiosis of Animals with Plant Microorganisms* (New York: John Wiley and Sons, 1965), 71.

26. Ibid., 74.

27. Joshua Lederberg, "Exobiology: Approaches to Life beyond the Earth," *Science* 132 (1960): 393–400.

28. See also Carl Sagan, Elliott C. Levinthal, and Joshua Lederberg, "Contamination of Mars," *Science* 159 (1967): 1191–96. While scientists such as Lederberg and Sagan were worried about possible contamination of Mars, other biologists, especially those concerned with so called germ-free research, worried about contaminating the moon.

29. "Report of the President," Rockefeller University Research Profiles, Spring 1990, 1–6, 5.

30. See, for example, S. Bradbury, *The Evolution of the Microscope* (Oxford: Pergamon Press, 1967).

31. See, for example, German Sims Woodhead, *Bacteria and Their Products* (London: Walter Scott, 1891), 24.

32. For an overview of such studies of mitochondria, see Lars Ernster and Gottfried Schatz, "Mitochondria: A Historical Review," *Journal of Cell Biology* 91 (1981): 227–55.

33. R. Y. Stanier and C. B. van Niel, "The Concept of a Bacterium," *Archiv für Mikrobiologie* 42 (1962): 17–35. Edouard Chatton, *Titres and Travaux Scientifiques* (Sète: Sottano, 1937).

34. Stanier and van Niel "The Concept of a Bacterium," 21.

35. See also P. Echlin, "The Relationship between Blue-Green Algae and Bacteria," *Biological Reviews* 40 (1965): 143–87.

36. See Ernster and Schatz, "Mitochondria," 227s–225s.

37. Stanier and van Niel, "The Concept of a Bacterium," 33.

38. R. Stanier, M. Douderoff, and E. Adelberg, *The Microbial World, 2nd ed. (Englewood Cliffs, N.J.: Prentice-Hall, 1963).*

39. For an account of the origins of genetic research on *Chlamydomonas* see Jan Sapp, *Where the Truth Lies. Franz Moewus and the Origins of Molecular Biology* (New York: Cambridge University Press, 1991).

40. See Ruth Sager, *Cytoplasmic Genes and Organelles* (New York: Academic Press, 1972). Nicholas Gillham, *Organelle Heredity* (New York: Raven Press, 1978).

41. Hans Ris and Walter Plaut, "Ultrastructure of DNA-Containing Areas in the Chloroplast of *Chlamydomonas,*" *Journal of Cell Biology* 13 (1962): 383–91, 388–90.

42. Sylvan Nass and Margit M. K. Nass, "Intramitochondrial Fibers with DNA Characteristics," *Journal of Cell Biology* 19 (1963): 613–28, 627.

43. R. Klein and A. Cronquist, "A Consideration of the Evolutionary and Taxonomic Significance of Some Biochemical, Micromorphological and Physiological Characteristics in the Thallophytes," *Quarterly Review of Biology* 42 (1967): 105–296, 167.

44. Ernest C. Pollard, "The Degree of Organization in the Bacterial Cell," in Katherine Brehme Warren, ed., *Formation and Fate of Cell Organelles. Symposia of the International Society for Cell Biology,* vol. 6 (New York: Academic Press, 1967), 291–303.

45. Aharon Gibor, "Inheritance of Cytoplasmic Organelles," in Katherine Brehme Warren, ed., *Formation and Fate of Cell Organelles. Symposia of the International Society for Cell Biology,* vol. 6 (New York: Academic Press, 1967), 305–16.

46. Ibid., 314.

47. Ruth Sager, "Mendelian and Non-Mendelian Heredity: A Reappraisal," *Proceedings of the Royal Society* 164 (1966): 290–97, 296.

48. William Bateson, "Address of the President of the British Association for the Advancement of Science," *Science* 40 (1914): 287–302, 293.

49. Voltaire, *Candide or Optomism,* trans. John Butt (London: Penguin Books, 1947), 20. For recent evolutionists, see S. J. Gould and R. C. Lewontin, "The Spandrels of San Marco and the Panglossian Paradigm: A Critique of the Adaptationist Programme," *Proceedings of the Royal Society of London B* 205 (1979): 581–98.

50. D. Halder, K. Freeman, and T. S. Work, "Biogenesis of Mitochondria," *Nature* 211 (1966): 9–12, 12.

51. Jostein Goksøyr, "Evolution of Eucaryotic Cells," *Nature* 214 (1967): 1167.

52. Lynn Sagan, "On the Origin of Mitosing Cells," *Journal of Theoretical Biology* 14 (1967): 225–74.

53. F. J. R. Taylor, "Implications and Extensions of the Serial Endosymbiosis Theory of the Origin of Eukaryotes," *Taxon* 23 (1974): 229–58.

54. Interview with Lynn Margulis, Bellagio, Italy, 29 June 1989.

55. J. Smith-Sonneborn and W. Plaut, "Evidence for the Presence of DNA in the Pellicle of *Paramecium,*" *Journal of Cell Science* 2 (1967): 225–34.

56. Chandler Fulton, "Centrioles," in J. Reinert and H. Ursprung, eds., *Origin and Continuity of Cell Organelles* (Berlin: Springer-Verlag, 1971), 170–220. John L. Hall, Zenta Ramanis, and David J. L. Luck, "Basal Body/Centriolar DNA: Molecular Genetic Studies in *Chlamydomonas,*" *Cell* 59 (1989): 121–32. Karl Johnson and Joel Rosenbaum, "Basal Bodies and DNA," *Trends in Cell Biology* 1 (1991): 145–49.

57. L. R. Cleveland and A. V. Grimstone, "The Fine Structure of the Flagellate *Mixotricha paradoxa* and Its Associated Micro-organisms," *Proceedings of the Royal Society B* 159 (1964): 668–86, 672.

58. Ibid., 681.

59. Sagan, "On the Origin of Mitosing Cells," 249.

60. Ibid., 231.

61. Ibid., 228.

62. Ibid., 247.

63. Ibid., 252.

64. Ibid., 270.

65. Lynn Margulis, *Origin of Eukaryotic Cells* (New Haven: Yale University Press, 1970), 65. R. H. Whittaker, "New Concepts of Kingdoms of Organisms," *Science* 163 (1969): 150–59.

66. Sagan, "On the Origin of Mitosing Cells," 270.

Chapter 12

1. R. Y. Stanier, "Some Aspects of the Biology of Cells an
Evolutionary Significance," in H. P. Charles and B. C. Knight, eds., *(*

Control in Prokaryotic Cells. Twentieth Symposium of the Society for General Microbiology (Cambridge: Cambridge University Press, 1970), 1–38, 31.

2. Karl Pearson, *The Grammar of Science* (London: Dent, 1937), 5.

3. A. Allsopp, "Phylogenetic Relationships of the Procaryota and the Origin of the Eucaryotic Cell," *New Phytologist* 68 (1969): 591–612, 604.

4. Ibid., 600.

5. Ibid., 606–7.

6. Ibid., 605.

7. Ibid.

8. Peter H. Raven, "A Multiple Origin for Plastids and Mitochondria," *Science* 169 (1970): 641–46, 645.

9. Ibid., 646.

10. John B. Hall, "The Nature of the Host in the Origin of the Eukaryotic Cell," *Journal of Theoretical Biology* 38 (1973): 413–18.

11. Rudolf A. Raff and Henry R. Mahler, "The Non-symbiotic Origin of Mitochondria," *Science* 177 (1972): 575–82, 575.

12. R. A. Raff and H. R. Mahler, "The Symbiont That Never Was: An Inquiry into the Evolutionary Origin of the Mitochondrion," *Symposia of the Society for Experimental Biology* 29 (1975): 41–92, 41.

13. Raff and Mahler, "The Non-symbiotic Origin," 576.

14. C. de Duve, "Origin of Mitochondria," *Science* 182 (1973): 85.

15. F. J. R. Taylor, "Implications and Extensions of the Serial Endosymbiosis Theory of the Origin of Eukaryotes," *Taxon* 23 (1974): 229–58, 245–46.

16. For recent evidence supporting Woese's suggestion, see Tom Fenchel and Catherine Bernard, "A Purple Protist," *Nature* 362 (1993): 300.

17. F. J. R. Taylor, "Autogenous Theories for the Origin of Eukaryotes," *Taxon* 25 (1976): 377–390, 378.

18. Ibid., 386.

19. Ibid.

20. Philip John and F. R. Whatley, "*Paracoccus denitrificans* and the Evolutionary Origin of the Mitochondrion," *Nature* 254 (1975): 495–498, 498.

21. Jeremy Pickett-Heaps, "The evolution of the Mitotic Apparatus: An Attempt at Comparative Ultrastructural Cytology in Dividing Plant Cells," *Cytobios* 3 (1969): 257–80.

22. Taylor, "Implications and Extensions," 251.

23. Jeremy Pickett-Heaps, "The Evolution of Mitosis and the Eukaryotic Condition," *Biosystems* 6 (1974): 37–48, 38–39.

24. T. Cavalier-Smith, "The Origin of Nuclei and of Eukaryotic Cells," *Nature* 256 (1975): 463–68, 467.

25. T. Cavalier-Smith, "Eukaryote Cell Evolution," in W. Geuter and B. Zimmer, eds., *Proceedings of the XIV International Botanical Congress* (Koenigstein: Koeltz Scientific Books, 1988), 203–23, 221.

26. J. T. O. Kirk and R. A. E. Tilney-Bassett, *The Plastids. Their Chemistry, Structure, Growth and Inheritance,* 2nd ed. (Amsterdam: Elsevier/North-Holland Biomedical Press, 1978), 247–48.

27. Ibid., 247.

28. Ibid., 248.

29. Thomas Uzell and Christina Spolsky, "Mitochondria and Plastids as Endosymbionts: A Revival of Special Creation?" *American Scientist* 62 (1974): 334–43, 342.

30. Ibid., 334.

31. Ibid.

32. Ibid., 343.

33. W. H. Woolhouse, Review of *The Plastids* by J. T. O. Kirk and R. A. E. Tilney-Bassett, *New Phytologist* 66 (1967): 832–33, 833.

34. Stanier, "Some Aspects of the Biology of Cells," 31.

35. Ibid., 24.

36. For Dobzhansky's views on the debate over the symbiosis theory of the eukaryotic organelles, see Theodosius Dobzhansky, Francisco J. Ayala, G. Ledyard Stebbins, and James W. Valentine, *Evolution* (San Francisco: W. H. Freeman, 1977), 383–88.

37. R. B. Roodyn and D. Wilkie, *The Biogenesis of Mitochondria* (London: Methuen, 1968), 58.

38. Uzzel and Spolsky, "Mitochondria and Plastids as Endosymbionts," 342.

39. Taylor, "Implications and Extensions," 247.

40. Lynn Margulis, "Symbiotic Theory of the Origin of Eukaryotic Organelles: Criteria for Proof," *Symposia of the Society for Experimental Biology* 29 (1975): 21–38, 21.

41. Ibid., 21.

42. Ibid., 35.

43. Lynn Margulis, *Symbiosis in Cell Evolution* (New York: W. H. Freeman, 1981).

44. Laurence Bogorad, "Evolution of Organelles and Eukaryotic Genomes," *Science* 188 (1975): 891–98, 891.

45. See C. R. Woese, "Archaebacteria," *Scientific American* 244 (1981): 98–122. C. R. Woese, O. Kundler, and M. L. Wheelis, "Towards a Natural System of Organisms: Proposal for the Domains Archaea, Bacteria and Eucarya," *Proceedings of the National Academy of Sciences* 87 (1990): 4576–79.

46. W. Ford Doolittle and Linda Bonen, "On the Prokaryotic Nature of the Red Algal Chloroplasts," *Proceedings of the National Academy of Sciences* 72 (1975): 2310–14; "Partial Sequences of 16s RNAs and the Phylogeny of Blue-Green Algae and Chloroplasts," *Nature* 261 (1976): 669–73.

47. L. B. Zablen, M. S. Kissle, C. R. Woese, and D. E. Buetow, "The Phylogenetic Origin of the Chloroplast and the Prokaryotic Nature of Its Ribosomal RNA," *Proceedings of the National Academy of Sciences* 72 (1975): 2418–22.

48. L. Bonen, R. S. Cunningham, M. W. Gray, and W. F. Doolittle, "Wheat Mitochondrial 18s Ribosomal RNA: Evidence for Its prokaryotic Nature," *Nucleic Acid Research* 4 (1977): 663–71.

49. D. F. Spencer, M. N. Schnare, and M. W. Gray, "Pronounced Structural Similarities between the Small Subunit Ribosomal RNA Genes of Wheat Mitochondria and *Escherichia coli*," *Proceedings of the National Academy of Sciences* 81 (1984): 493–97.

50. D. Yang, Y. Oyaizu, H. Oyaizu, G. J. Olsen, and C. R. Woese, "Mitochondrial Origins," *Proceedings of the National Academy of Sciences* 82 (1985): 4443–47.

51. W. Ford Doolittle, "Revolutionary Concepts in Cell Biology," *Trends in Biochemical Science* 5 (1980): 146–49.

52. M. W. Gray and W. F. Doolittle, "Has the Endosymbiont Hypothesis Been Proven?" *Microbiological Reviews* 46 (1982): 1–42.

53. Michael Gray, "The Endosymbiont Hypothesis Revisited," *International Review of Cytology* 141 (1992): 233–57.

54. Michael Gray, "The Bacterial Ancestry of Plastids and Mitochondria," *BioScience* 33 (1983): 693–99, 698.

55. See Gray, "The Endosymbiont Hypothesis Revisited."

56. See Lynn Margulis and Dorion Sagan, *Microcosmos: Four Billion*] *Evolution from Our Microbial Ancestors* (London: Allen and Unwin, 1987).

Chapter 13

1. Lynn Margulis, "Early life: The Microbes Have Priority," in William Irwin Thompson, ed., *Gaia. A Way of Knowing. Political Implications of the New Biology* (Great Barrington, Mass: Lindisfarne Press, 1987), 98–109, 109.

2. For an account of creationist movements from the nineteenth to the late twentieth century, see Ronald, L. Numbers, *The Creationists* (New York: Alfred A. Knopf, 1992).

3. See, for example, Niles Eldredge, *Unfinished Synthesis. Biological Hierarchies and Modern Evolutionary Thought* (New York: Oxford University Press, 1985).

4. See David Raup, *The Nemesis Affair. A Story of the Death of Dinosaurs and the Ways of Science* (New York: W. W. Norton, 1988).

5. Seymour S. Cohen, "Are/Were Mitochondria and Chloroplasts Microorganisms?" *American Scientist* 58 (1970): 281–89, 282.

6. Gunther S. Stent, *An Introduction to Molecular Genetics* (San Francisco: W. H. Freeman, 1971), 622.

7. Lewis Thomas, *The Lives of a Cell. Notes of a Biology Watcher* (New York: Viking Press, 1974), 71.

8. Ibid., 73–74.

9. Ibid., 76.

10. Ibid.

11. Ibid., 79.

12. See Lynn Margulis and Dorion Sagan, *Microcosmos. Four Billion Years of Evolution from Our Microbial Ancestors* (London: Allen and Unwin, 1987).

13. René Dubos, "Gaia and Creative Evolution," *Nature* 282 (1979): 459–62.

14. Thomas, *The Lives of a Cell,* 145.

15. Ibid., 148.

16. For an account of the origin of Lovelock's Gaia concept, see James E. Lovelock, "Hands Up for the Gaia Hypothesis," *Nature* 344 (1990): 100–102; and *The Ages of Gaia* (Oxford: Oxford University Press, 1989), 1–14.

17. Lovelock, "Hands Up for Gaia," 100.

18. James, Lovelock, *Gaia: A New Look at Life on Earth* (Oxford: Oxford University Press, 1979), 11.

19. Ibid., 73–77.

20. Ibid., 12.

21. Rachel Carson, *Silent Spring* (Boston: Houghton Mifflin, 1962).

22. Lovelock, *Gaia,* 107–8.

23. Ibid., 114.

24. Ibid., 116.

25. Ibid., 122.

26. Ibid., 117.

27. René Dubos, "Symbiosis between the Earth and Humankind," *Science* 193 (1976): 459–62.

28. Dubos, "Gaia and Creative Evolution," 154.

29. Ibid.

30. René Dubos, *The Wooing of Earth. New Perspectives on Man's Use of Nature* (New York: Charles Scribner's Sons, 1980).

31. See, for example, William Irwin Thompson, ed., *Gaia. A Way of Knowing.* (Great Barrington, Mass: Lindisfarne Press, 1981); Peter Bunyard and Edward Goldsmith, eds., *Gaia. The Thesis, the Mechanisms and the Implications* (Wadebridge, Cornwall: Quintrell, 1988); Peter Bunyard and Edward Goldsmith, eds., *Gaia and Evolution* (Bodwin, Cornwall: Abbey Press, 1989).

32. Lovelock, "Hands Up for the Gaia Hypothesis," 102.

33. Lynn Margulis, "Jim Lovelock's Gaia," in Peter Bunyard and Edward Goldsmith, eds., *Gaia. The Thesis, the Mechanisms and the Implications* (Wadebridge, Cornwall: Quintrell, 1988), 50–62, 50; "Gaia in Science," *Science* 259 (1993): 745. In her recent edition of her classic text, she writes, "Gaia is not an individual, it is an ecosystem." *Symbiosis in Cell Evolution* (New York: W. H. Freeman, 1993).

34. Lynn Margulis, "Biologists Can't Define Life," in Connie Barlow, ed., *From Gaia to Selfish Genes. Selected Writings in the Life Sciences* (Cambridge: MIT Press, 1991), 236–238, 237.

35. Lovelock, "Hands Up for the Gaia Hypothesis," 102.

36. John Postgate, "Gaia Gets Too Big for Her Boots," *New Scientist* 7 (1988): 60.

37. W. C. Allee, Alfred E. Emerson, Orlando Park, Thomas Park, and Karl P. Schmidt, *Principles of Animal Ecology* (Philadelphia: W. B. Saunders, 1949, rpt. 1961), 729.

38. Richard Dawkins, *The Extended Phenotype* (Oxford: W. H. Freeman, 1982), 236–37.

39. Ibid., 235.

40. W. Ford Doolittle, "Is Nature Really Motherly?" *The Co-Evolutionary Quarterly* 29 (1981): 58–65.

41. See Robert N. Brandon and Richard M. Burian, eds., *Genes, Organisms, Populations. Controversies over the Units of Selection* (Cambridge: MIT Press, 1984).

42. There are a vast number of books and articles on the sociobiology debate. I will mention here Arthur Caplan, ed., *The Sociobiology Debate* (New York: Harper and Row, 1978); Joe Crocker, "Sociobiology: The Capitalist Synthesis," *Radical Science Journal* 13 (1983): 55–71; W. R. Albury, "The Politics of Truth: A Social Interpretation of Scientific Knowledge, With an Appliction to the Case of Sociobiology," in Michael Ruse, ed., *Nature Animated* (Dordrecht: Reidel, 1983), 115–29; S. Rose, L. J. Kamin, and R. C. Lewontin, *Not in Our Genes. Biology and Human Nature* (Harmondsworth: Penguin, 1984); Philip Kitcher, *Vaulting Ambition. Sociobiology and the Quest for Human Nature* (Cambridge: MIT Press, 1987).

43. Robert Trivers, "The Evolution of Reciprocal Altruism," *Quarterly Review of Biology* 46 (1971): 35–39, 45–47. Reprinted in Arthur L. Caplan, ed., *The Sociobiological Debate* (New York: Harper and Row, 1978), 213–226, 220.

44. Richard Dawkins, *The Selfish Gene* (Oxford: Oxford University Press, 1976), 202.

45. Ibid., 3.

46. Ibid., 197.

47. Ibid.

48. Doolittle, "Is Nature Really Motherly," 60–61.

49. Dawkins, *The Extended Phenotype,* 237.

50. Robert Axelrod, *The Evolution of Cooperation* (New York: Basic Books, 1984), 89.

51. G. C. Williams, *Adapation and Natural Selection: A Critique of Some Current Evolutionary Thought* (Princeton: Princeton University Press, 1966). Robert May, *Stability and Complexity in Model Ecosystems* (Princeton: Princeton University Press, 1973).

52. R. May, "Mutualistic Interactions among Species", *Nature* 296 (1982): 803–4; I "A Test of Ideas about Mutualism," *Nature* 307 (1984): 410–11.

53. Axelrod, *The Evolution of Cooperation,* 103.

54. Ibid., 100.

55. Ibid., 105.

56. Ibid., 143.

57. Doolittle continued to emphasize this point ten years later; see Barlow *From Gaia to Selfish Genes,* 235–36.

58. Ibid., 235–36.

59. For a discussion of the anthropic principle in physics, see George Greenstein, *The Symbiotic Universe. Life and the Cosmos in Unity* (New York: Quill, 1988).

60. Doolittle in *Barlow, From Faia to Selfish Genes,* 33.

61. See Mary Beth Saffo, "Symbiosis and the Evolution of Mutualism: Lessons from the *Nephromyces*-Bacterial Ednosymbiosis in Molgulid Tunicates," in Lynn Margulis and René Fester, eds., *Symbiosis as a Source of Evolutionary Innovation* (Cambridge: MIT Press, 1991), 410–29.

62. D. C. Smith and A. E. Douglas, *The Biology of Symbiosis* (London: Edward Arnold, 1987), 262.

63. There are objections to this view of symbiosis. See, for example, R. A. Lewin, "Symbiosis Budgets: Profit and Loss Accounting," *Symbiosis* 4 (1987): 351–54.

64. For an account of the origin of these institutions, see Werner Schwemmler, "Symbiogenesis in Insects as a Model for Cell Differentiation, Morphogenesis, and Speciation," in Lynn Margulis and René Fester eds., *Symbiosis as a Source of Evolutionary Innovation* (Cambridge: MIT Press, 1991), 178–204, 189–91.

65. D. C. Smith, Foreward to Lynn Margulis and René Fester, eds., *Symbiosis as a Source of Evolutionary Innovation* (Cambridge: MIT Press, 1991), ix.

66. Lynn Margulis, "Words as Battle Cries—Symbiosis and the New Field of Endocytobiology," *Bioscience* 40 (1990): 673–77.

67. Ibid., 675.

68. Ibid.

69. See, for example, Lynn Margulis, "A Review: Genetic and Evolutionary Consequences of Symbiosis," *Experimental Parasitology* 39 (1976): 277–49. F. J. R. Taylor, "Symbioticism Revisited: A Discussion of the Evolutionary Impact of Intracellular Symbiosis," *Proceedings of the Royal Society of London B* 204 (1979): 267–86; "An Overview of the Status of Evolutionary Cell Symbiosis Theories," *Annals of the New York Academy of Sciences* 503 (1987): 1–17.

70. See S. Sonea and M. Panisett, *A New Bacteriology* (Boston: Jones and Bartlett, 1983).

71. See Lynda Goff, "Symbiosis, Interspecific Gene Transfer, and the Evolution of New Species: A Case Study in the Parasitic Red Algae," in Lynn Margulis and René Fester, eds., *Symbiosis as a Source of Evolutionary Innovation* (Cambridge: MIT Press, 1991), 341–63.

72. Paul Nardon and Anne-Marie Grenier, "Serial Endosymbiosis Theory and Weevil Evolution: The Role of Symbiosis," in Lynn Margulis and René Fester, eds., *Symbiosis as a Source of Evolutionary Innovation* (Cambridge: MIT Press, 1991), 153–169, 165.

73. Richard Law, "The Symbiotic Phenotype: Origins and Evolution," in Lynn Margulis and René Fester, eds., *Symbiosis as a Source of Evolutionary Innovation* (Cambridge: MIT Press, 1991), 57–71, 68.

Index